PROFESSIONAL PAPERS CORPS OF ENGINEERS.

NO. 14.

SIEGE ARTILLERY

IN THE

CAMPAIGNS AGAINST RICHMOND,

WITH

NOTES ON THE 15-INCH GUN,

INCLUDING

AN ALGEBRAIC ANALYSIS OF THE TRAJECTORY OF A SHOT IN ITS RICOCHETS UPON SMOOTH WATER.

ILLUSTRATED BY ACCURATE DRAWINGS OF A LARGE COLLECTION OF THE RIFLE PROJECTILES AND FUZES USED BY EACH ARMY IN VIRGINIA.

BY BVT. BRIG. GEN. HENRY L. ABBOT, U. S. ARMY,
MAJOR, CORPS OF ENGINEERS,
Late Bvt. Maj. Gen. Vols., com'dg Siege Artillery before Richmond.

NEW YORK:
D. VAN NOSTRAND, PUBLISHER.
1868.

THIS BOOK IS NO. 269
OF 1000 COPIES.

1986 LIMITED EDITION

Published by

Dean S. Thomas
P.O. Box 52
Arendtsville, Pa. 17303

The assistance of Mr. Wendell W. Lang, Jr. and Mr. William L. Smith is gratefully acknowledged and appreciated.

Printed in the United States of America

No. 14.

PROFESSIONAL PAPERS

OF THE

CORPS OF ENGINEERS UNITED STATES ARMY,

PUBLISHED BY AUTHORITY

OF

THE SECRETARY OF WAR.

HEADQUARTERS CORPS OF ENGINEERS.

ESSAYONS

1867.

TABLE OF CONTENTS.

LIST OF PLATES.

WOOD CUTS.

INTRODUCTION.

Scope of the Memoir. Record of the 1st Connecticut Artillery. Organization of the siege artillery brigade before Richmond. Artillery data collected.

The campaigns against Richmond were chiefly those of a large army manœuvring in the field, where siege artillery was of secondary importance. Still two large trains were brought into use, one in 1862 and one in 1864–'65, and two batteries of siege guns accompanied the army of the Potomac in its campaigns of 1863.

The rapid progress made of late in the science of artillery demands close attention from the Corps of Engineers. Indeed any facts bearing upon the capabilities, uses, and theory of modern ordnance possess an interest almost as great for engineer as for artillery officers. For these reasons I have devoted such time during the past year as my professional duties would allow, to preparing the following memoir, designed to place in a small space the most important results of the recent experience in Virginia. Incidentally, an analysis of the problem of ricochet firing upon water has been attempted, based upon certain data collected before my volunteer command was disbanded at the end of the war.

Having no personal experience with the train of 1862, except as an officer of engineers, I shall make no attempt to elaborate the reports of Colonel Tyler, commanding, and of his ordnance officer, Major Doull, both of which are added to this paper as appendix A. Reference should also be made to the detailed report of General Barry, chief of artillery, army of the Potomac, already published.*

For details concerning the two heavy batteries moving with the army, reference should be made to appendix B, containing a report of Captain Pratt, who commanded one of them.

---

*Report of the Engineer and Artillery Operations of the Army of the Potomac, from its organization to the close of the Peninsular Campaign.—*Barnard* and *Barry*, 1863.

To the train of 1864–'65, which was under my personal command, the following pages are chiefly applicable. Many valuable details concerning it will be found in the report of Major Hatfield, contained in appendix C.

The important batteries of siege guns in all these campaigns were served by the 1st regiment of Connecticut artillery, which was thus identified in a conspicuous manner with the history of the army of the Potomac. This paper is therefore in some sort a record of its labors, and especially of its contributions to the science of artillery.

RECORD OF FIRST CONNECTICUT ARTILLERY.

In May, 1861, this regiment was mustered into the United States service as infantry. On January 1st, 1862, it was changed to artillery. After a few months of drilling in the defences of Washington, it went into the field under Colonel Tyler to serve the siege train of 1862. It there took rank as one of the best disciplined and most efficient regiments in the army, and became imbued with a spirit of enthusiasm for the duties of its special arm.

In January, 1863, after the promotion of Colonel Tyler, his Excellency W. A. Buckingham, Governor of Connecticut, conferred upon me (then captain of engineers, United States army) the appointment of colonel of the regiment. From that date until its muster out of service in September, 1865, it remained under my command; constituting the basis of an artillery brigade which sometimes exceeded an aggregate of 3,500 men.

During 1863 the regiment was stationed in the defences of Washington, except companies B and M, which served the two heavy batteries already mentioned. They each consisted of four $4\frac{1}{2}$-inch ordnance guns, and were attached to the artillery reserve of the army of the Potomac, where they were under the immediate command of Brigadier General Hunt, chief of artillery.

The ten companies in the defences of Washington under Brigadier General Dé Russy, as division commander, and

Brigadier General Barry as chief of artillery of the department, had every facility for a thorough training in both the theory and practice of siege artillery.

At the opening of the campaign in 1864, the united regiment was ordered to more active service; and from that date to the capitulation of Lee the siege artillery brigade served all the heavy guns in front of Richmond, firing about 65,000 rounds, or over 1,200 tons of ammunition.

Subsequently it removed all the captured ordnance of a calibre sufficiently large to require any special professional skill in its handling.

A detachment with a siege train accompanied General Terry at the taking of Fort Fisher, North Carolina, and there had some rare experience in unloading heavy guns on an exposed ocean beach.

At an epoch when progress in the science of artillery is as rapid as at present, operations so extended should be turned to account in developing the theory of the arm. Inspired with this idea, and eager for the scientific reputation as well as for the efficiency of the regiment, its officers undertook and succeeded in collecting a series of records heretofore unattempted in actual warfare.

The present pages place these records in a convenient form for future reference, together with such practical deductions as are suggested by an experience of more than two and a half years in a responsible artillery command.

ORGANIZATION.

In order to decide what confidence shall be accorded to any series of observations, it is essential that the system under which they have been conducted shall be known. This is especially true in a case like the present, where a machine so extensive and complicated as a siege train is handled by volunteer troops. For this reason, as well as because the efficiency of artillery depends very much upon its organization, I deem it proper to give in some detail the composition of the command, with its interior arrangements and system of administration.

The organization of the 1st Connecticut artillery, in common with the other volunteer artillery regiments, was the following:

1 Colonel.
1 Lieutenant Colonel.
3 Majors.
1 Adjutant, (not extra lieutenant.)
1 Quartermaster, (not extra lieut.)
1 Surgeon.
2 Assistant Surgeons.
1 Chaplain.
1 Sergeant Major.
1 Quartermaster Sergeant.
1 Commissary Sergeant.
1 Hospital Steward.
2 Principal Musicians.

Twelve companies, each consisting of:

1 Captain.
2 First Lieutenants.
2 Second Lieutenants.
1 First Sergeant.
1 Quartermaster Sergeant.
6 Sergeants.
12 Corporals.
2 Musicians.
2 Artificers.
1 Wagoner.
122 Privates.

Making a company aggregate of 152, and a regimental aggregate of 1,839, officers and men. The regiment was armed with the Springfield rifled musket in addition to its artillery, and was thoroughly drilled in all infantry manœuvres.

The commissioned officers were appointed by the governor of the State, and, like the enlisted men, were mustered into the United States service by the United States mustering officers. They took rank from the date of muster, after which the governor had no military control over officers or men, the regiment being governed by the same rules and regulations as the regular United States troops.

No supervision was exercised by the United States government as to the qualifications or merits of these appointees. An officer having a commission from the governor was mustered into service, provided there was a legal vacancy; unless, as a punishment for previous misconduct, he had been disqualified from holding office. Examining boards were often convened to report on the qualifications of officers considered to be deficient, who were discharged if the report was unfavorable; but no such case ever occurred in this regiment. The whole responsibility for the selection of its officers thus rested upon the governor of Connecticut.

With a wise appreciation of the requirements of a good military organization, his excellency Governor Buckingham uniformly appointed its officers from the regiment, and never except upon the recommendation of its commanding officer. It is hardly possible to over-state the advantages conferred by this system. The *materiel*, like that of most of the volunteer regiments, was of an unusually good character; college graduates being by no means unknown to its ranks, while the majority were possessed of a good common-school education.

The commanding officer had thus in his power, by making just and judicious recommendations to the governor, to offer to the members of the regiment a career open to merit, uninfluenced by political or other favoritism; and to secure to himself the aid of a body of officers in every way qualified for their duties. This was naturally considered a matter of primary importance; and whenever a few vacancies had occurred the following system was adopted in selecting the nominees. Each captain was called upon for the name of his non-commissioned officer best entitled by acquirements, character, and faithful performance of duty to receive a commission. Field officers added those personally known to themselves to the list, which thus usually contained from twenty to thirty names. An examining board, consisting of all the field officers, including the commanding officer, was then convened, and the candidates were thoroughly examined in artillery and infantry. All the non-commissioned officers were habitually required to recite to their company officers in the text books treating of those branches whenever the duties of the regiment would allow; and these examinations were consequently thorough, exceeding half an hour to each candidate. After carefully comparing his own notes of the examination with the lists of the other field-officers, due weight being given to the character of the candidate for soldierly qualities and energetic performance of duty, the commanding officer submitted his recommendations to his excellency the governor.

At every promotion, up to captain inclusive, the senior in each grade passed a similar examination before the same board, so that if an unworthy appointment had been made no subsequent promotion would be obtained.

The power of assignment and of transfer among the different companies being vested entirely in the hands of the regimental commander, he was enabled to promote non-commissioned officers into companies where they were strangers to the enlisted men; to give to officers failing in one company an opportunity of benefiting in another by their past experience; and to see that each company contained officers well qualified for all varieties of duty. Great attention was always paid to this matter, which was judged to be of vital importance.

The character of the officers of the regiment can be inferred from the method of their appointment. The jarring factions so common to most volunteer regiments were nearly unknown to the 1st Connecticut artillery. Its officers were a body of men who appreciated the dignity of their position, and who were well educated in their duties. The regimental record shows how those duties were performed.*

It remains to explain the general system of administration adopted under the different circumstances in which the command was placed.

---

* The following is an extract from a letter addressed to me by Major General W. F. Smith, on August 20, 1864, when we happened to be needing recruits:

"During the time I commanded the 18th corps before Petersburg, I called heavily upon you for siege guns and mortars, and never before during the war have I witnessed such artillery practice as I saw with your regiment. The practicability of holding my position there after the 21st of June was due in a great measure to the skill displayed by your batteries. I trust every effort will be made to fill up a regiment which has not its equal in artillery firing, and which cannot be dispensed with without great injury to the service."

The following extract from a letter from General Barry, dated July 27, 1865, speaks for itself:

"As chief of artillery successively of the two principal armies of the United States during the four years of war now happily ended, I have enjoyed unusual opportunities for observation. You will on this account value my opinion, when I assure you that the 1st Connecticut artillery, in intelligence and the acquirements and services of its special arm, stands unrivalled in the armies of the United States."

In the defences of Washington the matter was one of great simplicity. Each major was placed in command of the contiguous forts garrisoned by his battalion. He receipted for all ordnance property; was responsible to the regimental commander for the discipline, instruction, and vigilance of his command; and all communications from his company commanders passed through his hands. It should be added that companies were never permanently assigned to battalions, but were placed under the different majors according to convenience and the necessities of the locality, an assignment rendered imperative by the length and nature of the lines.

In the siege of Richmond a somewhat similar system was adopted. Each major was placed in command of a convenient collection of batteries, irrespective of the number of companies serving them, or of their regiment. His headquarters were at some central locality, generally near those of the commander of the army corps on whose front his batteries were placed. From this officer, under certain restrictions imposed by the general-in-chief, he received orders as to actual firing; in all other respects his orders proceeded from brigade headquarters.

The lieutenant colonel (Colonel White) was usually stationed at brigade headquarters, performing the duties of inspector general, aiding the commanding officer as needed, and always prepared, from his intimate knowledge of the lines, to assume command of part, or of the whole, in case of necessity.

My own position was anomalous, owing to the fact that our guns extended along the fronts of both the army of the Potomac and the army of the James. During most of the siege I reported directly to the commanding general of the latter and to General Hunt, chief of artillery of the former, for the portions of the brigade serving with each army respectively. The train was organized at the request and upon a project drawn up by General Hunt, and would have been under his orders exclusively had it not been necessary to extend the use of the guns beyond the lines of the army of the Potomac.

The depot, under the charge of Major Hatfield, ordnance officer of the siege artillery, was established centrally at Broadway landing, on the Appomattox river, where three substantial wharves were built. The orders were to keep the material afloat; and no ammunition even was unloaded except to be put on the wagons. The vessels varied in number from a dozen to twenty, and consisted of schooners and barges of about one hundred and fifty tons each, and of a steam-tug for moving them and communicating with City Point. A permanent train of fifty government wagons, together with the four light artillery teams of company M, 1st Pennsylvania artillery, supplied the usual demands for transporting ordnance stores and guns. In emergencies, additional teams and wagons were furnished by General Ingalls, chief quartermaster of the armies in the field. A field telegraph connecting with the main lines gave direct communication with all the batteries, and greatly facilitated operations.

One or two companies were usually stationed at the depot, under the orders of Major Hatfield, to load material, perform guard duty, &c.; and when the army of the Potomac began to extend towards the left, a system of strong field works was constructed for the defence of the post against cavalry raids.

The supply of the batteries was difficult, owing to the fact that they covered a front of about seventeen miles in extent; and that, unlike field artillery, no caissons or limber chests accompanied the guns. All ammunition had to be kept in field magazines of limited capacity, and liable to dampness. Any accumulation being thus impracticable, the system of supply had to be very exact and prompt. The following was adopted; and, although over 1,200 tons of ammunition, hauled by wagon over bad roads an average distance of eight miles, were expended in the campaign, in no single instance did a battery fail to be supplied for ordinary or extraordinary demands; and in no case, except at the final evacuation, did a useless accumulation of ammunition occur.

Each battery commander forwarded daily to his major at reveille a detailed return of ammunition. These papers were collected by mounted orderlies from brigade head-quarters, and reached the depot about noon. Trains were at once fitted out to supply deficiencies below a certain number of rounds, (usually 100 per gun or mortar,) ordered to be kept on hand in the field magazines. These trains reported to the majors already informed by telegraph of their destination and time of starting, and were thus conducted under cover of the darkness to the proper batteries. In case of any sudden demand, a telegram from the major in charge at once notified the depot, and started a special train.

Responsibility for ordnance property was very simply settled for the siege train. The whole material remained charged to the ordnance officer; memorandum receipts, which were destroyed when the property was accounted for to him, being required from battery commanders. No time was thus wasted upon unnecessary papers, which the constant transfer of ordnance from one battery to another would have rendered very troublesome to company commanders, had the usual system been in force.

The men were well supplied with rations, clothing, &c., by the brigade quartermaster and commissary, Brevet Major Mason, who had a special train of seventeen wagons for the purpose.

The senior medical officer, Surgeon and Brevet Lieutenant Colonel Skinner, organized an independent field hospital at the depot, where milder cases could be well treated. On the approach of winter, two log houses, sixty by twenty feet, well roofed with shingles, were constructed without cost to the government. By avoiding the sending of light cases to general hospital. whence they seldom returned, the strength of the command was much increased. The assistant surgeons were assigned to the staffs of the majors, and did gallant service in the batteries.

The acting assistant adjutant general, Brevet Major Learned, efficiently transacted the complicated office duties of the command.

Such was the general administration of the brigade; and it will be noticed that the volunteer artillery organization was admirably adapted to the requirements of the siege train. Experience has suggested no change as upon the whole desirable, and Congress in the recent increase of the army has given almost identically this organization to each of the five regiments of artillery in the regular service.

DATA COLLECTED.

The subject of siege artillery may naturally be treated under the heads of mortars, of smooth-bore guns, and of rifled guns. Under each head will be comprised the records of the target practice in the defences of Washington, together with the greatly extended results of the actual firing before Richmond, where every effort to secure valuable data for future use was made. Careful men were specially detailed in each battery, to note the effect of every shot, fuze, &c. These records in their consolidated form furnish tables of practice never before attempted in actual service, and hence of interest to engineer and artillery officers. The results of experimental firing to test new inventions, and of a close scrutiny of the confederate ordnance, will also be added; together with notes upon the merits and uses of the different kinds of modern artillery, as developed by our experience in the field. A few remarks upon earthworks and siege batteries will close the paper.

CHAPTER I.---MORTARS.

Their use and history with the armies in Virginia. Remarks on our mortars, mortar-beds and platforms. Guns and howitzers as mortars. Notes on mortar shrapnel and on new projectiles, including hand-grenades. Precautions to be observed in the use of mortars. Present system of powder measures defective. View of the enemy from the battery essential. Expedients for pointing. Precision and range. Latter greatly influenced by form of chamber. New tables of ranges much needed. Practical test of our mortar fuzes. The growing importance of vertical fire, both with armies in the field and in sea-coast defence. Capture of Fort Fisher considered in this connection. A 20-inch mortar necessary.

After a little experience in campaigning in Virginia, both armies adopted the expedient of immediately intrenching themselves upon taking up a position. This was chiefly caused by the murderous precision of the rifled small-arms with which both armies were supplied, but was also very useful in covering the troops from horizontal artillery fire. The rapidity was surprising with which the "rifle pits," so called, could be thrown up by the aid of a few axes for felling timber, and shovels, or even bayonets or tin cups in cases of necessity, for moving dirt. After a few hours' work, the men lay secure in their trenches, indifferent alike to artillery and musketry fire. Vertical fire alone could severely annoy them.

Although at the siege of Yorktown we had placed heavy mortars in position, and had practically experienced the annoyance of receiving upon our approaches the constant fire of one 8-inch mortar from the confederate lines, neither belligerent availed itself during the subsequent campaigns of this species of artillery, prior to the advance of the army of the Potomac from the Rapidan, in the spring of 1864. In preparing for this campaign, General Hunt, chief of artillery, had procured eight Coehorn mortars to accompany the movement, and had included in the composition of my siege train many 10-inch and 8-inch siege mortars and Coehorns. It was a new arm to the troops, and excited much interest and attention from both officers and men. Owing to the peculiar character of the march from the Rapidan to Petersburg, but little use was made of the Coehorns; although they were placed in position and served on the lines at Cold Harbor. After the failure of the first assault upon the confederate position on the heights of Petersburg, the siege train was called into active service, and then began, for the first time in the experience of the armies operating in Virginia, a really heavy mortar fire.

Having noticed the effect upon the morale of our troops produced by the indifferent practice of the single 8-inch mortar from the confederate lines of Yorktown, I had paid great attention to training the gunners in the use of this arm, while in the defences of Washington. They were familiar with all the minor but essential details upon which the effect of vertical fire depends; and, as I learned at the time from deserters, and subsequently from confederate officers, the result of their sudden and unexpected opening on the Petersburg lines was appalling. Having no mortars wherewith to reply, and no bomb-proofs for cover, and yet being compelled by the proximity of the main lines (only two hundred yards distant in the nearest place) to keep their own fully manned in order to guard against an assault, the enemy suffered severely for the first few days, and the moral effect was extremely depressing. On one occasion which came under my personal observation, a confederate soldier was blown entirely over his parapet by the explosion of one of our shells ; and his body lay, the clothing consuming by fire, beyond the reach of his friends, who were deterred from approaching by our sharpshooters. To thus deprive an opponent of the accustomed protection of the trenches is well calculated to shake his nerves preparatory to an assault, or to retard or prevent the pushing of siege approaches.

As soon as the enemy could obtain mortars they placed them in position, and from that time until the evacuation the fire was frequent and severe on several points of the lines. Our expenditures amounted to over forty thousand rounds, and theirs were not much less. Having the benefit of previous training, our gunners retained the advantage, and the precision of their practice was justly admired.

Mortars were first introduced and multiplied upon the Petersburg front, with a view to preparing the way for an assault, and for keeping the artillery of the enemy quiet when it was delivered. This purpose they accomplished most effectually. When, after the Mine fiasco, the project of directly assaulting the confederate position was abandoned, the fire was maintained to keep down picket firing and to compel the silence of certain very annoying batteries, which, from the left bank of the Appomattox river, enfiladed the right of our line and caused much loss. Both parties ultimately constructed bomb-proofs, and remaining as much as possible under cover when the firing was going on, received little injury For this reason it was gradually discontinued on this part of the line.

At Dutch Gap, however, the confederates had an opportunity to reap the full advantage of vertical fire, and they continued it there until the canal was essentially completed, sinking one of the dredges and greatly harassing our working parties. They placed their mortars in sunken batteries, provided with good bomb-proof cover, on the low ground on the right bank of James river, in front of the canal. Our return fire was so heavy that they fell into the error of concealing these batteries behind clumps of trees, and thus lost, what is really essential to success, a good view of the target from the battery itself. The result was that their range was faulty, and their shells fell chiefly from forty to one hundred yards to the eastward of the canal, where they literally ploughed up the ground. This error enabled our working parties to continue the excavation with comparatively little loss, but suffering greatly from the harassing nature of the fire.

A second error, worse than the first, committed by the confederates at Dutch Gap, was in not sufficiently multiplying their fire. They never used more than four or five mortars, and these chiefly Coehorns. Not less than twenty, and these eight and ten inches in calibre, were required. I am confident that this number could have been placed in position in such a manner that no efforts on our part could have compelled their silence; and that well served, they would have effectually prevented the digging of the canal.

To check their fire as much as possible, 10-inch, 8-inch, and Coehorn mortars were used—advantage being taken of the high signal tower at Crow Nest to correct errors of direction—and also horizontal fire of shell and case shot from field guns, and, occasionally, from a 100-pounder rifle, to annoy the confederate gunners in watching the effect of their shot. To compel, by such means, resolute soldiers like these confederate artillerymen, to suspend fire from well-constructed mortar batteries was impossible; but the precision of their practice was so much impaired that the work on the canal could continue.

MORTARS, MORTAR BEDS, AND PLATFORMS.

The new model of the 8-inch siege mortar, 10-inch siege mortar, and 13-inch seacoast mortar, was exclusively used before Richmond, with the new wrought-iron bed.—(See wood cut at end of chapter for a rough sketch of it.) These calibres differ only in dimensions and strength, the principle of having the trunnions so placed as to give no preponderance, and the substitution of the elliptical for the gomer chamber, being adopted in all of them. The new is a very great improvement over the old model. The quoin is replaced by the elevating bar working from a fixed fulcrum upon a ratchet on the breech of the mortar, thus greatly facilitating the pointing of the piece; the wooden bolster, so constantly giving trouble by splitting, done away with entirely; the range of the mortar is materially increased by the change in the form of the chamber, as will soon be more fully explained; and last, but not least, the arrangement of the metal gives additional strength over that of the old model, where the mass of iron forming the trunnions at the base of the breech is liable to cause the destruction of the mortar by unequal expansion. I have never known one of the new model mortars to burst, and some of them in our batteries were fired over two thousand times without any perceptible enlargement of the vent or other sign of wear.

The Coehorn mortars and the 10-inch seacoast mortars supplied to the train were of the old model. That of the former does not seem to require essential modification; that of the latter, as just explained, is entirely out of date. It was more difficult to manipulate this piece than the new model 13-inch mortar of nearly three times its weight.

The following is the record of endurance for the few mortars which were fired the largest number of times. None were injured during the siege, except by the fire of the enemy.

10-inch	siege	mortar,	No.	14,	fired	538	times.	Apparently uninjured, the vents scarcely showing wear.
"	"		"	22,	"	524	"	
8-inch	"		"	20,	"	1,502	"	
"	"		"	24,	"	1,777	"	
"	"		"	25,	"	1,603	"	
"	"		"	26,	"	1,660	"	
"	"		"	32,	"	2,167	"	
"	"		"	36,	"	2,167	"	
24-pd'r	Coehorn	mortar	"	87,	"	1,114	"	
"	"		"	89,	"	1,127	"	
"	"		"	106,	"	1,123	"	
"	"		"	110,	"	1,168	"	

There are two improvements which ought to be introduced in all mortars, as our experience fully proved:

1. *The line of metal should be permanently marked upon the mortar.* As the external diameter is the same at the muzzle and breech, this line must be parallel to the axis whether the bed is level or not. The marking could be done, once for all, in the shop, and far more accurately than in the battery. The gunner's level and chalk could thus be thrown aside; a decided advantage in actual service, where keeping a battery supplied with such small articles is difficult.

2. Mortars *must be arranged for firing by friction primers.* This will always be done in actual service, both because it

is much simpler than the use of the slow match, and because it avoids the lighting up of the battery at night, which serves to direct the enemy's fire. Friction primers cannot be conveniently used with the present mortar and bed; first, because there is no convenient guide to keep the lanyard perpendicular to the vent, as must be done to prevent the primer from being drawn out; (this difficulty is enhanced by the worse than useless recess for priming;) and, second, because there is no cap to stop the metallic part of the primer from flying with great violence out of the vent and endangering the cannoniers.

The battery commanders obviated the first of these difficulties to some extent by attaching wires to the bed, but never succeeded in making an arrangement so convenient as could readily be devised for a permanency. The second defect could not be remedied. One of my officers lost his eye in consequence, and over a dozen of the officers and men were more or less severely cut.

Upon representing these defects to the Ordnance department an arrangement attached to the bed was designed to obviate them, but was not received until after the end of the siege, and was therefore not practically tried.

My idea would be to screw a small plate, say 3×2×0.2 inches, to the mortar, at the rear of the vent. This plate would support at its rear extremity a ring just large enough to admit the lanyard hook, and at its front a cap (like those used on the Whitworth breech-loaders,) to project over the vent high enough to admit of the introduction of a friction primer. This contrivance would be compact, would be easily removed when the mortar was transported, and would effectually remedy both difficulties. Some such invention should be provided for all mortars having sufficient weight to allow the use of the lanyard. The recess for priming at the top of the vent should be suppressed in all but the Coehorn mortar, in which at present, strange to say, it is not found.

The platforms supplied by the Ordnance department are

not sufficiently solid to endure much firing. We invariably had to place them upon large timber skids, bedded below the sleepers, and to anchor a heavy log at the ends of the deck plank.

Contrary to the usual theory, it was found necessary to use a platform with the Coehorn mortar, whenever any accuracy was required.

The great weight of the 13-inch mortar (17,000 pounds) renders it difficult to move, and some satisfactory experiments were made with a novel platform. An ordinary railroad platform car (eight wheels) was strengthened by additional beams tied strongly by iron rods, and was plated on top with iron. The mortar was placed upon this car, (top of mortar nine feet above track,) and run down on the Petersburg and City Point railroad, to a point near our lines, where a curve in the track afforded facilities for changing the plane of fire by advancing the car or drawing it back. The mortar, fired with fourteen pounds of powder, recoiled less than two feet on the car, which moved ten or twelve feet on the track. The effect of the charge was taken up without damage to the axles, even when the full allowance of twenty pounds of powder was used. This mortar, whose shell would crush and explode any ordinary field magazine, excited dread among the confederate gunners, and was effective in inducing their enfilading batteries on Chesterfield Heights to discontinue fire upon the right of our line. Its practice was excellent. At the battle of the Mine, as reported by three different observers stationed at different points, the explosion of one of its shells blew a confederate field gun and carriage above the parapet, at a range of about 3,600 yards. Of course with this platform the plane of fire must be nearly parallel to the track, or the mortar will be dismounted; but by placing the car on a curve a very considerable traverse can be secured without difficulty.

The advantages of vertical fire under certain circumstances are so great, that in May, 1864, a few experiments were made, under my direction, by Captain Osborne, 1st

Connecticut artillery, to test the light 12-pounder gun as a mortar. They indicated that it might be thus used when mortars could not be procured; but the expedient was never necessary in the siege of Richmond.

The carriage was dismounted and the gun reversed in its trunnion beds so as to point over the trail. Two parallel skids laid in a direction perpendicular to the parapet and separated by the proper interval, supported the axle at such a height as to allow the breech to be depressed between them sufficiently to give an angle of elevation of 45° to the piece. The breech rested on a kind of quoin, and it was found necessary to tie the trail down with the prolonge when the gun was fired. With five ounces of powder, giving a range of about 1600 yards, the strain upon the carriage was inappreciable. Solid shot (weighing 12.25 pounds) were used; and with charges of four ounces and less, the sabot was not detached until striking the ground. With very small charges the sabot passed below the vent, but the friction primer always ignited the powder. These experiments were made in great haste, at the request of General Hunt, chief of artillery, army of the Potomac, only three days before starting for the field, and no very extended trial was practicable.

The following table exhibits the results :

Vertical fire from the light 12-pounder.

Kind of ordnance.	Mortar powder.	Elevation.	Range.	Remarks.
	Oz.	°	*Yds.*	
12-pounder, light...	1	45	125	Projectile, solid shot, weighing 12.25 lbs.
Do ....do....	2	45	496	
Do ....do....	3	45	617	Penetration, 2½ feet.
Do ....do....	3½	45	1 009	Approximate.
Do ....do....	4	45	1 200	Approximate.
Do ....do....	5	45	1 600	

Similar experiments with an 8-inch howitzer were conducted by Lieutenant Colonel Haskin, in charge of the defences north of the Potomac, in October, 1863. The following are the results obtained by him and communicated to me in April, 1864. The howitzer was mounted in the manner above described.

Vertical fire from the 8-inch siege howitzer.

Charge. (Mortar powder.)	Range.			Remarks.
	New model howitzer.	Old model howitzer.	Difference.	
Ounces.	*Yards.*	*Yards.*	*Yards.*	
4	314	300	14	Projectile, 8-inch shell.
8	620	553	67	
12	1 082	Lost.	....	
16	1 440	1 332	108	
20	1 925	1 695	230	

The variation in range between the two models, of which the bores differ only in the form of the chamber, (elliptical and gomer,) will be noticed with surprise. It accords with my own experiments, soon to be given, with the new and old model mortars. The absolute value of the range shows that a considerable economy in powder would result from using the howitzer instead of the mortar, but in service this advantage would be more than balanced by the greater inconvenience of loading and pointing. It is, however, a fact worth remembering that vertical fire, in cases of necessity, can be obtained readily and effectively from guns and howitzers.

PROJECTILES.

Knowing that a vertical fire of spherical case shot had been tried in Belgium with a view to dispensing with stone mortars, I applied early in 1863 to Major Benton, commanding Washington arsenal, to prepare some projectiles for me in the usual manner, for experimental purposes. He suggested that the expedient be tried of filling the 10-inch shell with 12-pounder canister shot and adding the bursting charge loose. This I did in October, 1863; the first time, probably, that spherical case shot were ever fired from a mortar in this country.

The firing was at Fort Scott, in the defences of Washington, south of the Potomac, the new model 10-inch siege mortar being used. The target was in a valley fifty yards below the mortar and eight hundred and fifty yards distant. The projectile was the ordinary 10-inch mortar shell with twenty-seven of the balls of a 12-pounder canister (thirty-eight filled the shell) inserted through the fuze-hole, and a bursting charge of 2.5 pounds of powder added on top of them. The shell weighed ninety pounds and each ball 0.43 pounds, making the total weight one hundred and four pounds. A charge of one pound six ounces of mortar powder gave a range of eight hundred yards, with a time of flight of thirteen seconds. By placing observers at different stations to notice the points at which the shell was projected upon the distant hills, at its explosion, a close estimate of its height above the ground was secured.

The following table exhibits the results of the experiments:

Shrapnel from mortars.

No. of shot.	Charge of mortar.	Length of fuze.	Burst above target.	Remarks.
	Lb. oz.	*Sec.*	*Yards.*	
1	1 6	11.0	190	The fragments of shell mostly struck from 200 to 300 yards around target; found 64 balls well distributed within 40 yards of target; they penetrated about 3 or 4 inches into grassy turf. Only one small fragment of shell within this circle. Shot No. 2 fell short and should not be counted. Firing was then resumed.
2	1 5	12.0	50	
3	1 6	12.0	100	
4	1 6	12.5	100	
5	1 6	13.5	50	
6	1 6	14.0	Ground.	40 yards short; 20 balls blown out of hole lying near it.
7	1 6	13.5	Ground.	70 yards short; 9 balls blown out of hole lying near it.
8	1 6	13.0	2	66 yards short; found 22 shots in circle of 2 yards diameter, and 15 fragments of shell within 10 yards; all outside the balls.
9	1 6.5	13.0	45	50 yards short; found 23 shot in circle of 40 yards diameter, and one fragment of shell only within that distance.
10	1 6.6	13.5	7	45 yards short; found 27 shots in circle of 5 yards diameter, and 8 fragments of shell within circle of 15 yards diamater.

It was concluded from these experiments that when the shell burst, the balls fell in a cone about 30° at the vertex, while the fragments scattered very much more. The balls had, at this range, ample force to kill, penetrating from three to seven inches into a turf, where, when thrown by a man with his whole force, they entered less than one inch. Indeed, a little computation will show that the velocity at impact must have exceeded two hundred feet per second, which, with a projectile weighing nearly half a pound, supplies ample living force to disable man or horse. Of course, if the range were very short the requisite velocity would not be acquired, as it depends essentially upon that of the shell at the instant of bursting.

The fact that the force of the bursting charge is expended in fracturing the shell, and does not materially scatter the balls contained in it, obviates for this kind of projectile the great cause of failure in mortar fire against troops, viz: that if the shell is burst over the point occupied by the enemy, the fragments scatter so widely as to render the position nearly a safe one, unless the shell is *near the ground.* The uncertainty of fuzes renders this height a matter of practical difficulty to control, especially as the fragments of such shells as bury themselves before exploding do no damage whatever. The spherical case shot throws its balls evenly over a limited circular space, not exceeding in diameter its height above the ground at the instant of explosion, and hence must be far more effective than the shell in retarding the progress of works of siege, or in silencing batteries. It virtually extends the range of the stone mortar to that of the ordinary shell.

In accordance with these views this projectile was employed in the battle of Petersburg Mine, where General Hunt's orders for the artillery were to use every exertion to keep quiet the batteries of the enemy bearing upon the point of assault. A battery of ten 10-inch mortars was placed near the subsequent location of Fort Rice, and directed its fire, at a range of eight hundred yards, upon a salient battery of the confederates, from which we anticipated much trouble. Not a shot was fired from this battery after its range was obtained, and I was informed by a confederate officer that the gunners found it impossible to remain by their guns and endure the shower of balls falling from shells bursting about once in thirty seconds over the battery.

This same projectile was occasionally used during the rest of the siege, but the gunners conceived a prejudice against it from a few premature explosions near the muzzle of the mortar, which they attributed to the concussion acting upon the iron balls and loose powder inside the shell. No casualties resulted from these explosions, which I was never fully satisfied did not occur from the loosening of the wooden fuze plug that at long ranges necessarily extended down among the balls. A plug and paper fuze, like those supplied with the Coehorn shell, would obviate this difficulty; but it doubtless would be better to prepare a regular spherical case shot, in which the balls are held in place by sulphur. The confederate method of filling spherical case shot with iron balls might well be used in preparing such ammunition, viz : to have two holes in the shell, through one of which, subsequently closed by brass or lead, the balls and sulphur are introduced, while the cavity for the powder is preserved by a plug inserted through the regular fuze hole.

This projectile is not proposed for shells of a less diameter than ten inches, (or possibly eight inches with lead balls,) as smaller calibres are deficient in interior space. The balls should weigh nearly half a pound each to be effective. The expedient tried by the confederates in imitation of our projectiles, (and occasionally used by our own men,) of mixing musket bullets with the bursting charge of small shells, was entirely futile from want of weight in the bullets.

Their attempt to use small iron balls in their 12-pounder spherical case shot was a failure, whether fired from guns or mortars. In frequent instances, cannoniers were struck by these balls without injury.

Two patent shells were tested in the batteries. The

Pevey shell consisted of a single casting, making two thin concentric shells connected by studs. The space between was filled by small iron balls of the size of musket bullets, (or incendiary composition if desired,) inserted through a hole in the outer shell, afterwards closed by lead. The shell was solid at the fuze hole, which connected with the hollow of the inner shell, where was placed the bursting charge. Several hundred of these shells, of calibres 8-inch and 10-inch, made to weigh, when loaded with iron bullets, the same as ordinary shells of those calibres, were received for trial. Carefully tested, over water, and over a dusty plain, they appeared to separate into fully double the number of fragments usual to shells of these calibres, and to have proportionally greater effect. They were frequently used in the batteries, and always without premature fracture, even when fired with heavy charges. A 10-inch shell was fired from a seacoast mortar with a charge of seven pounds, without injury. For firing at troops I regard them as less effective than good spherical case, but decidedly an improvement upon the common shell. The space between the concentric shells is necessarily too contracted to receive iron balls sufficiently large to be of any utility.

The other patent tested was McIntyre's repeating shell, which consists of concentric shells so arranged that the explosion of the outer one by the ordinary means shall ignite a fuze, which shortly after explodes the second shell, and thus ignites the fuze of the inner shell. To insure prompt ignition should the projectile strike the ground before bursting, Mr. McIntyre used with his exterior shell a peculiar fuze, shown by Fig. 23, Plate I. It consists of an ordinary time fuze, containing within its composition a hollow brass cylinder *a*, imbedded in plaster of Paris *b*. This cylinder is open at both ends, the top being closed by a brass stopper *c*, kept in place by the plaster and fuze composition. The latter being quickly consumed, the plaster fails to hold the stopper at striking, and the flame is instantly admitted to the shell. This fuze gave favorable results on trial.

Only a few samples (calibres 10-inch, triple bursting, and 24-pounder, double bursting) were furnished, and these arrived just previous to the end of the siege. The shell, therefore, was never fully tested in the field. I am informed that, at a previous trial made by the 1st Connecticut artillery after the evacuation of Yorktown, the shell developed a great penetrating power by its successive explosions in the ground, thus indicating that it would be useful to explode field magazines. The samples fired in 1865 were thrown into the confederate lines, which prevented this property from being noted. The experiments were made in my absence by Major Brooker, 1st Connecticut artillery, with the following results:

Practice with McIntyre shell from mortars.

No. of shot.	Kind of projectile.	Charge.	Length of fuze.	Time of first explosion.	No. of explosions.	Remarks.
	Shells.	*Lbs.*	*Sec.*	*Sec.*		
1	10-inch.	1.75	20	18.2	2	Both explosions in the ground.
2	10-inch.	1.75	14	16	2	1st explosion in air, 2d in ground.
3	10-inch.	1.75	12	12	1	Burst in air.
4	10-inch.	1.75	10	....	2	1st burst at muzzle, 2d in air.
5	10-inch.	1.75	10	17	1	Burst in ground.
6	10-inch.	2.	10	14	3	Burst twice in air, once in ground.
7	10-inch.	2.50	12	....	3	Burst first at muzzle, twice in air.
	Shells.	*Oz.*	*Sec.*	*Sec.*		
1	24-pdr.	6.50	12	12.0	1	Burst in ground.
2	24-pdr.	6.50	11.5	11.5	2	Burst once in air, once in ground.
3	24-pdr.	6.75	11.5	11.5	2	Burst twice in air.
4	24-pdr.	6.75	11.5	11.5	2	Burst twice in air.
5	24-pdr.	6.75	11.5	11.5	2	Burst twice in air.
6	24-pdr.	6.75	11.5	11.5	2	Burst twice in air.
7	24-pdr.	6.75	11.5	11.5	2	Burst twice in air.
8	24-pdr.	6.75	11.5	11.5	2	Burst twice in air.
9	24-pdr.	6.75	11.7	11.7	2	Burst once in air, once in ground.

These shells were forwarded to me by General Dyer, Chief of Ordnance, with the following memorandum of experiments made by Major Benton, of that department:

The average weight of the repeating shell, loaded, was eighteen pounds three ounces. The average weight of common Coehorn mortar shells, seventeen pounds eleven ounces.

"The first shell was fired from a 24-pounder Coehorn mortar, with four ounces of powder and 9.5 second fuze. The first explosion took place at about five seconds after firing, and the second about one second afterwards. The range was about four hundred yards. Pieces found.

"2. Coehorn mortar; four ounces powder; 12-second fuze; exploded twice; pieces found.

"3. Coehorn mortar; four ounces powder; 12-second fuze; exploded twice; pieces found.

"4. Coehorn mortar; six ounces powder. Only one explosion was heard. No pieces found.

"5. This shot was fired from a 24-pounder field howitzer; 4-second fuze; $2^{\circ}.25$ elevation. The first explosion took place as the shell struck the ground. It then ricocheted and exploded.

"6. 24-pounder field howitzer; 5-second fuze; $2^{\circ}.5$ elevation. This shot ricocheted, but no explosion was heard.

"These projectiles are exceedingly simple in their construction, and weigh but little more than the ordinary 24-pounder shell, and they have the marked advantage of producing two explosions instead of one."

There were two improvements in mortar shells introduced by the confederates, which, in my judgment, should be adopted in our service.

The first consists in giving the interior of the shell a polyhedral form instead of that of a sphere concentric with the exterior. This was occasionally used with higher calibres, and extensively with the 12-pounder shells, both for guns and for a Coehorn mortar of that calibre, (made like their 24-pounder Coehorn mortar of iron.) These shells split in fragments on the bounding lines of the interior surface, and thus make it possible to supply the maximum number of pieces consistent with proper weight. Our shells frequently split into very few fragments, some of which are too large and others too small. I have often seen fragments, especially of the larger calibres, consisting of half the shell.

The confederates used two different patterns for the interior polyhedron, both indicated on Plate I. The first (Fig. 1) consists of a regular dodecahedron, of which the five top pieces are wanting on the drawing; the second, (Fig. 2,) of an upper and lower pentagon connected by ten equal trapezoids, of which the six top pieces are wanting on the drawing. In both cases the shell is designed to break into twelve equal parts. In the first pattern this is perfectly accomplished; in the second, the iron being much thicker on the joints, which, regarding the pentagonal pieces as marking the poles, may be described as equatorial, than on those following meridian lines, (as 0.8 is to 0.5 inches,) the trapezoidal pieces often adhere to each other by twos; thus giving only seven fragments to the shell. The first pattern is therefore decidedly the better.

The second confederate improvement consists in supplying their Coehorn mortar shells with ears. This is far better than our clumsy system of strapping with tin. No fixed ammunition should ever be supplied for this mortar, as was occasionally done, because it makes the shell dangerous to store and handle, especially when used under fire. Unfilled, the shells can be piled outside the magazine, thus economizing valuable space, and can always be prepared as fast as required by the cannoniers. My men preferred the unfilled shells, even unstrapped and without ears, to the fixed ammunition.

The lines lay so near each other at certain points on the Petersburg front as to render sudden night assaults probable. Requisition was accordingly made upon the Ordnance department for a few light balls of 8-inch calibre, which were supplied from West Point. They were tested experimentally as follows:

The mortar was of the new model. The projectile weighed 12.25 pounds. When the canvas over the fuze end was cut, the ball uniformly ignited and burned well for about ten minutes. The range and time of flight were somewhat reduced by a moderate wind from the front. According to the usual rule (one-twenty-fifth the weight of the ball) a charge of eight ounces of powder, giving a range of about three hundred and fifty yards, could be used; but it will be noticed that a charge of six ounces, giving a range of two hundred and fifty-five yards, was the maximum which the ball would endure without bursting open from the concussion of the fall. Perhaps, by diminishing the elevation, a little larger charge might be used ; but the increased range, rendered necessary by the improvements of modern warfare, can hardly be attained with these projectiles—especially with *fire* balls, which ought to have sufficient strength to reach and light up the enemy's lines.

Practice with light balls.

No.	Charge.	Elevation.	Time of flight.	Time of burning.	Range.	Remarks.
	Lbs. oz.	°	*Seconds.*	*Min.sec.*	*Yds.*	
1	0 2	45	4.5	11 00	70	Cut and burned.
2	0 3	45	5	....	106	Not cut; not burned.
3	0 4	45	5	....	133	Cut and burned.
4	0 5	45	7	....	200	Cut and burned.
5	0 6	45	8	....	255	Cut and burned.
6	0 7	45	10	....	310	Cut; burst on striking.
7	0 8	45	10	....	350	Cut and burned; burst on striking.
8	1 0	45	10	....	415	Cut and burned; burst on striking.
9	1 0	45	11	....	400	Not cut; not burned.
10	1 8	45	15	....	475	Not cut; burned; burst on striking.
11	1 8	45	15	0 40	565	Cut and burned; burst on striking.

The lines between the armies in front of Richmond were at no point so close together as to require the use of hand grenades. An ingenious pattern, ignited by the act of throwing, was, however, offered to me in December, 1864, by its inventor, Mr. J. S. Adams. The grenade consisted of the ordinary 6-pounder spherical shell—any other shell could be used if required—with a fuze of peculiar construction. The fuze (Fig. 22, Plate I) consisted of a leaden plug containing a common short time fuze *a*; this was ignited by a friction primer *b*, inserted in a cavity by the side of the time fuze, and held firmly in place by an iron disk *c*, secured by four leaden shoulders pressed down upon its top, (one of which, marked *d*, is shown in elevation in the drawing,) and pierced by a central hole *e*, to allow the escape of the gas. The wire of the primer passed through a second hole in this disk and was then bent sharply over so as to be entirely contained within the plug. An ordinary lanyard, with the end attached to the wrist, could be hooked to this primer and ignition resulted from the act of throwing. To preserve the fuze from dampness, a paper cap covered with shellac varnish was placed over the primer in the cavity marked *f*; before firing this was torn off by pulling a piece of tape arranged for the purpose. This grenade can be thrown by hand about 100 feet, and scatters its fragments nearly 200 yards. It possesses the merits of employing the time fuze, the moral effect of which exceeds that of the percussion, and of being destroyed if not ignited, so that it cannot be thrown back by the enemy. It would have received the preference over Ketchum's, or any other kind known to me, had the siege operations demanded the use of hand grenades.

PRECAUTIONS TO BE OBSERVED IN THE USE OF MORTARS.

To accurately direct vertical fire undoubtedly requires more professional skill than any other kind of artillery practice: First, because in horizontal fire the projectile passes nearly parallel to the ground, and if in the proper vertical plane has many chances of hitting from ricochet or otherwise; the mortar shell with its high trajectory has but one chance—it hits or misses; hence more absolute precision is necessary with the mortar than with the gun, to give equal results with the two arms. Second, because the greater time of flight of the mortar shell increases the disturbing influences of the wind, of eccentricity, &c., and renders it more difficult to secure exactness in the fuze. Third, because the gun admits of being directly sighted at the object, while the mortar must be pointed by a comparatively imperfect and indirect method.

There are only two means of obviating these difficulties : first, by multiplying the number of pieces designed to accomplish a certain effect—thus ten mortars will produce at least one hundred times the moral effect of one ; and second, by giving the most exact attention to every detail upon which the precision of fire depends.

Thus, for the platform, it *must be* both level and stable. I have heard Coehorn mortars abused because an accurate fire at a range of six hundred yards could not be attained by firing them upon the natural surface of the ground.

For the carriage and mortar, the requirements necessary to accuracy are fulfilled by our new model, except that the line of metal should be permanently marked on the mortar. This is often inaccurately done in the battery, for want of an exact level; as well as from the rough unturned surface of the mortar, which of itself precludes great nicety in a chalk line, especially with the ear in its present position.

For the ammunition, the first essential is that the shells should be of exactly uniform weight, and the powder of exactly uniform quality. The first may be attained by adding gravel or bullets to the lighter shells until all are brought to the same weight, a condition rarely fulfilled by the shells as issued. A platform balance for this purpose should be supplied to permanent water batteries. In the field, where great accuracy happens to be required, an upright, supporting on its top the centre of a horizontal rod loaded at one end by one of the heaviest shells, is a simple expedient. The second condition may be attained by pouring all the powder designed for the day's firing upon some clean surface and thoroughly mixing it.

For weighing the charge, the system in use in our service (measuring it in powder measures of different sizes with their interior surfaces ungraduated) is radically wrong; because the amount of powder which they will contain depends upon whether it is loose or compact, *i. e.* upon how it is introduced; because the measures are inexact (I have rarely tested a set of which the smaller and larger measures

3

would correspond, and they sometimes differ twelve per cent. among themselves;) and lastly because the want of interior graduation prevents any accurate measurement unless the full capacity of the measure happens to be required. The weight, not the volume, of the charge is to be determined, and in sea-coast batteries an accurate *balance only* should be used. In the field, where extreme precision is not generally so essential, a measure made upon the system adopted by the confederates would be a great improvement upon ours. Their measure (Fig. 3, Plate I) consisted of a hollow copper cylinder containing as a bottom a sliding copper cylinder closed at each end and graduated to ounces. The latter, which was made to fit so tightly as to retain its place by friction, could be so adjusted as to cause the measure to exactly contain the desired charge, and thus permit a straight-edge to be always used.

To secure the ignition of the fuze requires special care. In the mortars of small calibre this is accomplished by placing the shell so as to bring the fuze near the top of the bore. In larger mortars a less dangerous method is to lay a train of dry powder from the top of the shell to the fuze, and another where the fuze would strike the bottom of the bore in rolling out, both made to remain in place by wetting the iron. Wooden fuzes should be driven gently so as not to loosen the composition, and *should not be sunned.* They should be sawed accurately in a fuze box, and not by the eye.

Great care and judgment and much personal skill are required in pointing mortars. The idea is too prevalent that the battery can be located in any covered position, a view of the target being quite unimportant. While, in the absence of any better site, such a location may be preferable to none, no accurate firing can reasonably be expected from it. A distant observer reporting the effects of his shots to the gunner, gives him too vague an idea of their character to be of much service. He must see them himself to make accurate firing. For this reason, *mortar bat-*

teries should be located where good views of the enemy may be had, even at the expense of extra labor. Such at least has been my invariable experience.

The method of pointing with the cord and plummet, laid down in the tactics, gives good results, except that in the vicinity of sharpshooters the stakes cannot be placed on the parapet. Our gunners had many devices of their own for this, and, indeed, for the whole method of pointing. To avoid sharpshooters, they sometimes bored two holes about five feet apart in a board and inserted upright sticks or wires ; this apparatus was laid on the superior slope of the parapet. By bringing the two wires into a vertical plane passing through the target, and moving the board from side to side, the proper positions could be found for the peg on the interior slope and for the rear stake, to locate the mortar in the centre of the platform. Generally, the rear stake was replaced by a permanent, low, single-barred fence, extending along the rear of the platform, and having upon its rail notches indicating the places of the cord for the different hostile batteries. Oftentimes the cord was discarded, and the mortar pointed by suspending the plummet line upon this fence at the proper notch, and bringing the line of metal of the mortar into the vertical plane passing through the plummet line, and the permanent peg on the interior slope—the gunner standing well to the rear and stooping for the purpose. This is a simpler method than that laid down in the tactics, and in a high wind decidedly preferable, as the plummet remains more steady, being inserted if necessary in a pail of water.

The use of a wooden-handled steel scraper, made in the shape of a hoe with a double edge, curvature 6.5 inches, was found to materially shorten the time required to serve the 13-inch mortar when the usual iron scraper was used.

PRECISION AND RANGE OF MORTAR FIRE.

A definite idea of the actual precision of fire to be expected from artillery is of the first importance. This is not to be obtained from the results of firing at arsenals, where the guns are served by men generally more intelligent, and possessed of far more experience than the average of gunners even among the best of artillery troops ; and who moreover have the advantage of using new ammunition, with every facility for obtaining good results. It is perhaps for these reasons that no experimental tables have been prepared to show the probable accuracy of fire with the different artillery in use. The want of such tables having been felt in my own branch of the service, the limited firing allowed for target practice in the defences of Washington was conducted so as to throw light upon this subject. The gun targets were placed on steep hillsides, with the distance from the battery exactly fixed by triangulation. Men were detailed to plant a marked stake at each point of impact, and after the firing was over an accurate survey with tape-line and level was made, from which the point of passing the plane of the target could be deduced. In like manner mortar targets were placed on open ground, the shells were marked, and after the firing, were dug up and exactly located by survey. In the battery the gunners were frequently changed ; the ammunition was treated with all the care which could be used in the field, and the firing was conducted in every kind of weather which allowed the target to be clearly distinguished. The tables which show the aggregate results of this firing, although in some cases deduced from rather too few shots, have at least the merit of being based upon accurate data.

Such experiments, of course, do not determine the maximum accuracy which can be obtained, but they do give a practical idea of the firing of well-instructed troops. For mortars, this is shown in the following table, the elevation being forty-five degrees. Had the Atlanta been lying at the target with her bows directed toward Fort Richardson, distant half a mile, she would have been struck fourteen times out of seventy-five shots from the 10-inch mortar, fired, it will be remembered, as well in high winds as during calm weather.

Precision of mortar fire.

Mortars.	Fort.	Range.	Total number of shots.	No. of shells falling within the following distances from the target.							Mean impact from target.
				10 yds.	20 yds.	30 yds.	40 yds.	50 yds.	60 yds.	80 yds.	
		Yards.									*Yds.*
10-inch siege........	Richardson	856	75	4	15	25	44	55	64	68	37
10-inch siege........	Barnard...	875	65	3	15	23	37	43	52	61	40
8-inch siege........	Ward......	800	50	4	8	11	18	25	30	43	48
24-pounder Coehorn.	Richardson	856	40	1	3	6	7	8	13	17	......
24-pounder Coehorn.	Barnard...	875	23	1	1	1	6	8	10	12	......
24-pounder Coehorn.	Ward......	800	15	....	....	2	3	4	7	10	......

From this table it appears that with the 10-inch siege mortar, at a range of half a mile, the probable distance from the point of impact to the centre of the target is forty yards, and that about six-tenths of the shells will fall within this radius. With the 8-inch mortar, these numbers are fifty yards and five-tenths respectively. With the 24-pounder Coehorn mortar (for which this range is decidedly too great for good firing) about half the shells will fall within eighty yards of the centre of the target. It will be remembered that our sea-going iron-clads are about one hundred yards in length, and that with mortars the accuracy of the fire increases with the weight of the shell.

One very curious fact was discovered by this target practice, namely, that the new model 10-inch siege mortar has a materially increased range over the old model. They differ only in the form of the chamber, a longitudinal section of which in the new model is of elliptical section, semi-axes 7.5 and 5 inches respectively. In the old model the gomer chamber is used—length, 5 inches; superior diameter 7.6 inches; inferior diameter, 5 inches. The result is, that with an empty mortar the shell enters four inches further, and with one containing the maximum charge of four pounds three inches further, into the bore in the new than in the old model. In the former, the shell touches the powder for charges exceeding one pound; in the latter, it only touches with the maximum charge of four pounds.

The difference in range between the two models was discovered by our firing in October, 1863. It was at once reported, and in December of that year the following experiments were made with proof powder supplied for the purpose by the Ordnance department. Every precaution was taken, both in the measurements and in the selection of still days to secure accurate results. The projectile was the shell unfilled, weighing ninety pounds; the elevation was forty-five degrees.

Comparative ranges of new and old 10-inch siege mortars.

Charge.	Range, Mean of 5 shots.		Difference.	Time of flight, Mean of 5 shots.		Difference.
	Old model.	New model.		Old model.	New model.	
Lbs.	*Yards.*	*Yards.*	*Yards.*	*Seconds.*	*Seconds.*	*Seconds.*
0.5	123	189	66	......	6.4	......
1.0	276	545	269	6.9	10.4	3.5
1.5	522	854	332	......	14.2	......
2.0	774	1122	348	11.5	17.2	5.7
2.5	1144	1410	266	14.6	18.4	3.8
3.0	1466	1676	210	17.5	19.8	2.3
3.5	1811	1848	37	......	20.9	......
4.0	2028	2064	36	21.5	21.9	0.4

At the time when these experiments were made there were no new model 8-inch siege mortars under my control. It was subsequently found, in front of Petersburg, that a similar difference existed for that calibre, and of so great an amount as to render the published lists of ranges worthless with the new model. A correct table, using our service powder, was accordingly deduced for this mortar by experimental firing in September, 1864. No mortar of the old model being tested at the same time, the usual table for its range (Griffith's) is used in the following comparison. The table of ranges in the heavy artillery tactics indicates still greater differences. The projectile is the shell, weighing forty-five pounds ; the elevation is forty-five degrees.

Comparative range of new and old 8-inch siege mortars.

Charge.		New model.			Old model.		Difference.	
		No. of shots.	Range.	Time.	Range.	Time.	Range.	Time.
Lbs.	*Oz.*	....	*Yards.*	*Seconds.*	*Yards.*	*Seconds.*	*Yards.*	*Seconds.*
0	8	5	360	8.0	......	......	......	......
0	12	5	703	12.6	......	......	......	......
1	0	4	1082	15.1	750	12.2	332	2.9
1	4	....	1412	17.0	1100	14.0	312	3.0
1	8	4	1741	18.8	......	......	......	......
1	12	....	1985	20.0	......	......	......	......
2	0	1	2225	....	......	......	......	......

These results and the deductions to be made from the experiments of General Haskin with the 8-inch siege howitzer, already detailed, suggest the inquiry whether an economy of both powder and weight of mortar may not be secured by slightly lengthening the bore. It may be that this is not desirable for field service, as a reduction in charge would increase the difficulty of graduating it.

In this connection it may be well to call attention to the fact that careful experiments are required to perfect the tables of ranges now in use, before mortars can be satisfactorily served. Thus, for the old 10-inch sea-coast mortar which has been in our service for many years, the artillery officer can nowhere find any further information as to its range and time of flight than the bare announcement that with ten pounds of powder these quantities are respectively 4,250 yards and thirty-six seconds. Practice with this mortar is thus reduced to simple conjecture. Three of them were in position on James river, where no accurate measurement of range could be made. By a comparative analysis of their firing, and that of the 100-pounder Parrott, at certain batteries of the enemy, checked by computation, (range in feet equal to sixteen times the square of the time of flight,) I made the following rough table for my gunners. It should, of course, be replaced by an accurate table:

Approximate table of ranges for old model (1841) 10-*inch mortar.*

Charge.		Projectile.	Elevation.	Range.	Time of flight.	Remarks.
Lbs.	*Oz.*		°	*Yards.*	*Seconds.*	
3	4	Shell.	45	1950	19.0	
3	12	Shell.	45	2350	20.0	
4	0	Shell.	45	2500	22.2	
6	8	Shell.	45	3600	26.5	
10	0	Shell.	45	4250	36. (?)	(Printed range,) time of flight believed to be excessive.

For the new model 10-inch sea-coast mortar we have no information whatever as to range or time of flight, and yet these mortars are to be found in many of our forts.

Besides good tables for an elevation of 45°, the ranges of our sea-coast mortars should be accurately determined at 60°, and even at higher elevations, because it may become necessary to sacrifice, in some degree, accuracy of fire in order to obtain increased force of impact. For this reason, also, solid shot should replace shells when attempts are made to sink iron-clad vessels, and proper tables of ranges should be prepared accordingly.

It remains to give the results of the actual mortar firing before Richmond. So far as this tests the quality of fuzes in use, the results are given in the following table, which explains itself. The precision of the fire will be vouched for by the army and by those who viewed the shell-marked batteries of the enemy after the evacuation.

Practical test of mortar fuzes.

Kind of fuze.	Number used.	Uncertain.	Number tested.	Burned well.	Burned variable.	Did not burn.	Percentage serviceable.
13-inch mortar, wooden	218	54	164	128	18	18	0.78
10-inch mortar, wooden	2712	1114	1598	1284	256	58	0.80
8-inch mortar, wooden	13942	1893	12049	10625	891	533	0.88
Coehorn mortar, paper	15763	2543	13220	12001	584	625	0·90

THE GROWING IMPORTANCE OF VERTICAL FIRE.

From the difficulty in securing accuracy in vertical fire a prejudice against it, except for shelling a large area, exists, even among artillery officers. In my opinion, this prejudice is as unreasonable as would be the condemnation of a theodolite, because, in the hands of a careless or ignorant surveyor, it gives inaccurate work. It is admitted that mortar fire must fail if, as at the siege of Pulaski,* and at many other places, the artillery troops are untrained in the use of the arm. This kind of fire, above any other, requires thorough training. As already stated, there are many apparently trivial precautions to be taken, and there is much individual skill in pointing to be attained. These secured, the mortar battery becomes the most terrible on the lines of an army.† There is much that is appalling, and, if long continued, painfully harassing to the imagination, in enduring a skilfully directed mortar fire. The dull, distant report, the long interval of expectation, the at first imagined then steadily increasing whistle of the shell, the explosion and the hurtle of the fragments, carrying confusion in every direction, the consciousness that the trusted parapet yields no protection, and that the only security is in first constructing and then bur-

---

* "During both days of the bombardment the wind, which blew from right to left, was extremely unfavorable for mortar firing. This, in connection with the fact that the gunners had never before fired a piece, and had been drilled only ten days, accounts in some degree for the loss of so many shells from the mortars."—*Siege of Pulaski, Gillmore.*

† I am informed by Major General J. G. Foster, who was chief of engineers at the first bombardment of Fort Sumter, that the confederates had about twenty 10-inch mortars, with ranges varying from 1,000 to 1,900 yards. After about six or eight hours' firing they got the range of the fort perfectly, dropping two out of three shells within the work, and bursting them finely at the level of the upper tier. This fire was very galling, and was the chief cause of the barbette guns remaining unserved.

Major General Q. A. Gillmore expresses, on page 128 of his Engineer and Artillery Operations against Charleston, the following views: "The special defence of Fort Wagner was faulty in two particulars, viz: * * * * Second, curved fire was not used enough. The armament of the work contained but two mortars (one 8-inch, and one 10-inch.) These, when earnestly served, caused the most serious delay in the progress of our work, and on one occasion suspended it entirely."

rowing in unwholesome bomb-proofs—all these influences combined produce an effect upon troops which has but to be witnessed to induce a high respect for so terrible an engine of destruction.

In my opinion, Coehorn mortar batteries should be added to the reserve trains of armies in the field. A single government wagon would easily carry a 24-pounder Coehorn mortar, with 100 rounds of ammunition complete In other words, the transportation required would only be about one-half that of ordinary field artillery. I also think that, like the confederates, we should introduce a 12-pounder Coehorn mortar into our service, which, with 200 rounds of ammunition complete, could be carried on a government wagon over any roads where light artillery could follow. Its ammunition being identical in calibre with that of the light 12-pounder, could always be readily supplied from the general ammunition trains in case of necessity. For practice against troops, the 12-pounder Coehorn is decidedly more deadly than the 24-pounder; as its shell, when the fuze burns too slowly, does not bury itself on striking, and the fragments thus scatter widely.

If the mortar well served is useful in land operations, it is rapidly becoming *essential* to harbor defence. With an armored vessel, like our latest class of monitors, possessed of sufficient speed to escape the impact of rams, plated with fifteen inches of iron on the turrets, and with essentially fourteen inches of iron on the sides properly backed, it is necessary to find some means of attacking its only vulnerable points, namely, its bottom and its deck. The former may be reached by the torpedo ; the latter can, in general, be assailed only by the mortar. The importance of vertical fire in our water batteries is thus greatly increased by the recent developments in naval warfare, and every effort must hereafter be directed towards securing the maximum accuracy of fire.

These views as to the importance of mortar fire at iron-clad vessels are not entirely theoretical. At Fort Fisher,

North Carolina, the absence of this fire enabled Admiral Porter to execute a very simple and effective plan for assisting the land forces. He anchored his iron-clads and a few gunboats within easy range of the work, out of the traverse of many of the water-bearing guns, and deliberately fired at the guns and obstacles to assault on the land front; while the unarmored vessels of the fleet lying at a safer distance kept up so furious a random fire from their heavy ordnance that the confederate artillery men could not serve their barbette batteries, of which the armament of the fort exclusively consisted. Had there been twenty-five or thirty 13-inch mortars securely placed behind the high sand parapets and traverses, and served by gunners well drilled in their use, the iron-clads could neither have maintained their position, nor secured the precision of fire, to which almost exclusively the injury to the defences of the land front was due. Indeed, the captured letter-book of the commander of the garrison, Colonel Lamb, shows that he had too late awakened to the necessity of relying upon this kind of ordnance, for he wrote just before the starting of the final expedition against the work that he had been recently studying "Sir Howard Douglas on Gunnery," and had come to the conclusion that mortars ought to be supplied to him.

It is not designed to attribute the capture of Fort Fisher entirely to the want of vertical fire in its defence. Besides the enormous disproportion in guns between the fleet and fort indicated by the tables in appendix C, there were fatal errors in the work, such as the neglect to remove the forest and to occupy the sand-hill behind which our troops sheltered themselves within assaulting distance; the omission to provide an interior retrenchment or cavalier to flank the sand traverses on the land front; the neglect to provide proper obstacles in front of the work which its necessary want of a ditch rendered doubly important; and lastly, the folly of placing heavy guns on barbette carriages for land defence instead of light field guns, which could be kept under cover of the parapet out of all danger from the fire of the fleet until the moment of assault had come. The remedy of any of these defects, or even perhaps a vigorous charge by the troops in front of our lines of countervallation, while our assault was in progress, might possibly have caused its failure; but that many of the land-bearing guns were disabled and that the stockading in front of the work was greatly injured, (essential elements in the plan of attack,) is to be attributed to the uninterrupted and accurate firing of the iron-clad fleet, which, in my judgment, could not have been maintained had a sufficient number of heavy mortars been well served against the vessels.

To recapitulate, the fatal damage to the works was done by the iron-clads. Thirty 13-inch mortars could have kept up a discharge of one shell every ten seconds, any one of which striking a monitor on her deck would have penetrated* it. No horizontal fire—we had no mortars available, and under no circumstances are mortars afloat to be much dreaded—could have prevented brave men from serving these mortars behind a sand parapet like that of Fort Fisher, (twenty feet high.) Hence, if such mortars had been properly served, the iron-clads would not have been anchored within a range which commanded the accuracy of fire necessary to seriously damage the work from the unstable decks of a vessel.

That mortar fire will hereafter be an important element in sea-coast defence must, it is thought, be admitted. To

* From a careful study of the reports of the officers commanding the iron-clad vessels and the gunboats at the capture of Fort Fisher, (see report of the Secretary of the Navy for 1865) it appears that the average distance of their position from the fort was at least 1,200 yards, and probably more. Captain Noble's formula (see Report of Experiments of Ordnance Select Committee, 1866) indicates that a projectile thirteen inches in diameter, and weighing two hundred pounds, would require only four hundred feet striking velocity to penetrate a solid iron plate two inches thick; which would probably offer a greater resistance than the decks of the iron-clads opposed to Fort Fisher. This velocity could readily have been obtained.

General Duane, Lieutenant Colonel of Engineers, informs me that a 10-inch mortar shell filled with sand and fired from Fort McAlister at one of our monitors from a distance of about a mile, just penetrated her deck.

be effective *many mortars* of the *largest calibre* must be used. The calibre is a very important consideration, because at usual ranges ordinary mortar projectiles do not possess sufficient living force to perform the work required. The decks of our latest class of monitors (Kalamazoo class) are plated with three inches of iron supported by a substantial backing of timber; and the tendency has been constantly toward an increase in this protection. To penetrate such a deck, at usual ranges, would require larger mortars than any at present in our forts. Captain Noble's formula (Report of Experiments of Ordnance Select Committee, 1866) indicates that a 20-inch solid shot weighing 1,000 pounds would require four hundred and fifty feet striking velocity to penetrate a solid, unbacked plate four inches thick. It is therefore evident that mortars of at least this calibre must be introduced into our service, especially as the principal objection to them—their unwieldy character—has been greatly lessened by the recent improvements in mortar beds.

13-inch mortar battery (No. 4) before Yorktown.

CHAPTER II.---SMOOTH-BORE GUNS.

Their present uses. A reserve train for defence of redoubts sometimes advantageous. Notes on projectiles. Precision of fire with troops. The 15-inch gun. Experimental firing to determine its ricochet trajectory. Analysis of the data collected and resulting formulæ. Later experiments with 8-inch and 24-pounder guns to test these formulæ. Discussion of the general problem of loss of useful ricochet trajectory, resulting from increasing the height of the battery above the water level.

The era of smooth-bore siege trains has passed away. Except some small howitzers for defending redoubts, there were no smooth-bores in the train before Richmond other than a few 8-inch siege howitzers, brought because their ricochet fire and large interior capacity of shell might possibly be needed. This failing to be the case, they were not used against the enemy.

For defending positions against assault, however, no artillery can be more efficient than the 32-pounder or 24-pounder field howitzer. The former (equal in diameter of bore to the 100-pounder rifled gun) throws very large case shot and canister, and from its light weight may be kept out of sight and danger until the assault is delivered, when it can suddenly be run into battery and served with murderous effect. This was shown on June 2, 1865, when the 22d South Carolina regiment, Colonel Dantzler commanding, made a determined assault upon the small advanced redoubt Dutton, on our Bermuda front. This work was garrisoned by one company of the 1st Connecticut artillery, Captain Pride commanding, with a few dismounted cavalry supports. It was then incomplete, but was armed with two 32-pounder and one 24-pounder brass howitzers. Only two of them could be brought to bear on the assaulting party, but so rapid a canister fire was maintained as to repulse the column with severe loss, the colonel himself being among the killed. So demoralized was his command, that a lieutenant and twenty-two enlisted men surrendered rather than attempt to retreat under the fire. They were marched into our main lines by a detachment of the cannoniers.

It is a matter for consideration whether a large army, remaining, like the forces in Virginia, for long periods of time in intrenched positions, could not with advantage have at the depot a reserve train of these howitzers, served by a regiment of foot artillery, and ready to be brought forward by quartermaster mule trains as needed. When not serving their guns, the greater part of the regiment would act as guards for the reserve artillery; would be specially charged with the construction of magazines, gabions, fascines, &c., for battery uses; and would always be ready to accompany assaulting columns, in order to use without delay any captured artillery upon the retreating enemy*—a service which artillery troops alone can perform to advantage. By this plan, the demoralization of the light artillery, resulting from keeping them for months behind parapets separated from their horses and with no opportunity for drill, would be avoided; and a body of skilled troops for constructing, defending, and assaulting field-works would be secured.

---

* This duty was performed by the artillery with signal success at the assault of the lines of Petersburg by the 9th Corps on April 2, 1865. A detachment of four officers and one hundred men from the 1st Connecticut artillery, commanded by First Lieutenant W. H. Rogers, was detailed for the purpose. It was divided into three platoons, each commanded by a second lieutenant. Each platoon was divided into three gun detachments, consisting of ten men and a chief of piece. All the detachments were armed with their muskets, and provided with lanyards, primers, fuzes, priming wires, &c., necessary to the prompt serving of captured artillery.

This command joined the assaulting column near battery No. 20, and entered among the first the enemy's works. They instantly began to serve four captured light 12-pounder guns upon the retreating masses of the enemy. Two more light 12-pounder guns were moved by them across the work, under a heavy fire, and within half an hour were also opened upon the enemy.

These six guns were served most gallantly all day and during the night. About four hundred rounds captured with the pieces were expended, and a like amount in addition, which was carried by hand from our lines. The men not required to serve the pieces used their muskets effectively, expending all their own ammunition and much more taken from the prisoners and from the dead and wounded. They captured about fifteen prisoners, and turned them over to be sent to the rear. Much praise was justly given to Lieutenant Rogers, and to his officers and men, for their gallant conduct, which contributed to the success of the charge, and greatly to the repulse of the desperate assaults made by the enemy to retake the captured works.

PROJECTILES.

The patterns of our smooth-bore guns and howitzers, with the details of their ammunition, require no notice here.

As already explained, the confederates gave a polyhedral interior to many of their smooth-bore shells. For fuzes they used the Bormann and time, the latter with wooden or copper plugs. Their case shot were filled with iron balls instead of lead, as in our service. These balls and the sulphur were introduced through a second hole, the chamber for the bursting charge being preserved during the operation by inserting a rod through the fuze hole to the opposite side of the shell. The second holes were subsequently closed by a plug of lead, iron, or copper. Our experience proved that these iron balls possessed too little weight to be effective.

Major General Warren, commanding 5th Corps, caused a collection to be made along his lines in front of Petersburg, with a view to discover the details of the confederate ammunition. This he turned over to me, and the following is a classification of the smooth-bore samples:

Confederate smooth-bore projectiles.

Projectiles.		Number of samples collected.		
		6-pounder.	12-pounder.	24-pounder.
Shells	Wooden plug	1	9	5
	Copper plug	....	20	....
	Bormann plug	5	27	....
	Total	6	56	5
Spherical case, copper plug, iron balls	2d plug, lead	....	57	2
	2d plug, iron	....	16	....
	2d plug, copper	....	2	1
	No 2d plug	....	1	....
	Total	....	76	3
Solid shot		....	31	....
Total		6	163	8

I often noticed strange fragments of spherical shell lying in our batteries. Thus, with both 8-inch and 12-pounder shells, a fragment broken exactly on concentric inner and outer circles—inner diameter a little over an inch—was common.* It would seem that this must be caused by the indirect influence of the fuze-hole upon the bottom of the shell at the instant of bursting, but in what manner it would be difficult to explain. Samples of these fragments were sent to the artillery museum at West Point.

The confederates experimented somewhat with elongated projectiles for smooth-bore guns. Five patterns are shown on Plate I, the most peculiar of which are the three varieties of the Maury projectile of chilled iron for 10-inch and 8-inch Columbiads, (Figs. 4, 5, and 6.) I was informed in Richmond that these were tested, and with a favorable result, by the confederate ordnance department.

PRECISION OF FIRE OF SMOOTH-BORE GUNS.

As already stated, the regular drill practice allowed for the instruction of the 1st Connecticut artillery, when stationed in the defences of Washington, was so directed as to give an idea of the probable accuracy to be expected from the fire of good troops with the different guns in the field-works around that city. The following table exhibits the results of this practice for smooth-bore guns, made up from careful surveys:

Precision of fire of smooth-bore guns.

Ordnance.	Fort.	Range.	Total number of shots.	Number of shots within the following distances from the centre of target.						Mean impact from centre of target.
				5 ft.	10 ft.	20 ft.	30 ft.	40 ft.	50 ft.	
		Yds.								Ft.
8-inch S. C. howitzer	Barnard	980	55	2	15	30	44	51	55	20
8-inch siege howitzer	Ward	850	50	3	11	28	39	42	45	23
32-pdr. S. C. gun	Barnard	1030	30	3	5	15	25	29	30	20
32-pdr. S. C. gun	Richardson	950	34	0	4	15	24	28	.....	28
24-pdr. siege gun	Ward	850	29	7	16	23	27	28	.....	15
24-pdr. siege gun	Barnard	1030	30	4	7	23	26	30	.....	17
24-pdr. siege gun	Richardson	950	38	5	12	27	32	35	.....	20
24-pdr. field howitzer	Barnard	980	20	0	1	8	15	18	19	26
24-pdr. field howitzer	Richardson	950	18	0	0	3	9	10	12	39

*Brevet Major A. H. Burnham, Corps of Engineers, informs me that he noticed similar fragments of 15-inch shell fired by our fleet at the siege of Fort Morgan; but the inner diameter was larger, being not less than six inches.

One curious accident attending this practice is worthy of record. In firing from Fort Richardson, at a range of 950 yards a 32-pounder solid shot passed through the target and entered the hillside behind. A 24-pounder solid shot passed through the same hole and struck the first shot *exactly*. The former was entirely uninjured; the latter was split into several pieces in planes passing through the end of the trajectory (considered straight) and all radiating from the point of percussion, while a circle 2.5 inches in diameter was indented to the exact curvature of the 32-pounder. This shot is preserved in the Engineer department at Washington.

The penetration of a 24-pounder solid shot in a sound white-oak stump, at a range of eight hundred and fifty yards, was eight inches.

EXPERIMENTS WITH THE FIFTEEN-INCH GUN.

After the termination of hostilities, the 1st Connecticut artillery was ordered to the defences of Washington for a short time before muster out of service. It thus came under the command of Brigadier General Haskin as division commander. He was engaged upon some computations to correct and improve the published tables of ranges, and desired some very accurate experimental firing with the 15-inch gun, for checking the values of certain constants in the formulæ. This work was assigned to my brigade. Two forts within its limits contained these guns; one, Fort Foote, at an elevation of one hundred and three feet, and the other, Battery Rogers, at an elevation of thirty-six feet above the usual water level of the Potomac. Officers of the Corps of Engineers have long desired to know definitely how much the advantages of ricochet fire are sacrificed when the axis of the gun is raised at different heights above the water surface. The practice was accordingly conducted with a view to throw all possible light upon both of these important points. In fact, I fired a few of the shots at the request of the Chief of Engineers, and at elevations proposed by him, specially to test the matter of ricochet.

To secure the desired data, it was necessary, first, to fix with precision the exact point of as many impacts as possible at the different angles of fire; and second, to conduct the practice with every care to secure uniformity in the ammunition.

The first condition was fulfilled by covering the region of fire with a net-work of carefully measured triangles, so that three plane-tables could be exactly located, with respect to each other and to the gun, at such points as should give the best practicable angles for all points of probable impact. This was done with a good sextant (radius seven inches) and theodolite, by Major Hatfield and myself. All the angles of each triangle were measured, and the work was carefully computed, so that no error appreciable in a base for artillery firing could exist. Figs. 16, 17 and 18, Plate I, show to the eye the relative positions of the plane-tables and gun at the different firings. Accurate observers were placed at the instruments, which were of good quality. Their notes, when plotted on a large scale, proved conclusively, by the small triangles of intersection, that each impact was fixed with an average discrepancy of less than ten yards. The last few impacts at low elevations were usually so close together that it was impossible to locate them exactly; in fact, the ball almost appeared to roll on the surface of the water. The two observers in front of the gun, however, attempted to count the full number of them in each case. These numbers are given in the tables.

The second condition was fulfilled by the following precautions:

The entire amount of powder for each day's practice was poured out on the magazine floor, thoroughly mixed, and re-weighed into cartridges. It was of good quality.

The elevations were checked both by the elevating arc on the breech of the piece, and by a metallic quadrant in the bore.

The shells were weighed and filled with sand to the exact weight of 344.5 pounds for the first twelve shots, and 344 pounds for the remainder. Originally, they varied among themselves six or eight pounds, and in one instance fourteen pounds, from the mean.

The practice each day was begun by firing a blank cartridge. The chassis rails were cleaned and sanded at each discharge, the clamps being placed about six inches from the top carriage. The slope of the rail was three degrees.

The time of flight was fixed by two watches, usually stop-watches, and the mean adapted.

To form an approximate idea of the height attained by the column of spray thrown up at the first impact, a rod, graduated to feet and tenths, was placed vertically in the battery. The end of a tape-line was secured to this rod at the height of the mouth, the twenty-feet mark placed between the teeth, and the line straightened. The eye thus uniformly brought to the same place, read on the rod the apparent elevation of the bottom and top of the column of spray. The difference between these readings gave a side in the vertical triangle, whose base was twenty feet, proportional to the height of spray in the larger vertical triangle, whose base was the distance to the point of impact. The length of the latter being determined by the plane-tables, the height of the spray could readily be computed. Of course, these determinations were rough, but sufficiently accurate to possess interest, considering the variable nature of the object measured, which at top was nothing but a jet of vapor, much influenced by the wind.

Nearly calm days were selected for the firing, when the water was quite smooth. It was shallow, with grass growing in it near the lowest gun, (36 feet high) and the first and second impact at low elevations threw up muddy water; the rest, pure water.

Large fragments of sabot were thrown a measured distance of one hundred and sixty-one yards, in front of the highest gun, (103 feet high,) showing that sabots could not be used in one barbette battery firing over another.

The records of this practice with the 15-inch gun are, for convenience in size, presented in two tables, and are so classified as to exhibit the data relative to ricochet firing by themselves. To obtain all details respecting any shot, both tables must be consulted, the numbering, of course, being the same in both.

Practice with the 15-*inch gun.*

(TABLE No. 1.)

Number of shot.	Powder.		Elevation referred to horizon.	No. of impacts.		1st impact.			Last impact.	Recoil of gun.
	Kind.	Weight.		1st observer.	2d observer.	Time.	Range.	Height of spray.	Range.	
		Lbs.	°			*Sec.*	*Yds.*	*Ft.*	*Yds.*	*Ft.*
(*a*) 1	No. 5	40	2	19 +	23	....	955	....	3240	8.8
2	...do...	40	2	13 +	17	....	1040	....	3290	4.2
3	...do...	40	2	24	19	....	1100	....	3230	4.5
4	...do...	40	2	21	16	....	1105	....	3260	5.2
5	...do...	40	3	10	9	....	1340	....	2530	4.9
6	...do...	40	3	11	10	....	1375	....	2615	4.9
7	...do...	40	3	12	11	....	1350	....	2655	4.8
8	...do...	40	4	7	6	....	1580	....	2320	4.9
9	...do...	40	4	8	......	....	1565	....	2270	4.5
10	...do...	40	5	2	2	....	1770	....	1860	4.4
11	...do...	40	5	2	2	....	1855	....	1890	4.9
12	...do...	40	5	2	2	....	1880	....	1960	4.7
13	...do...	40	6	1	......	7.5	2100	185	2100	5.9
14	Mam.	40	6	1	......	6.7	1950	220	1950	5.0
15	No. 5	40	6	1	......	7.2	2025	305	2025	5.6
16	Mam.	40	6	1	......	7.0	1975	210	1975	5.6
17	No. 5	40	10	1	......	11.0	2810	....	2810	5.2
18	Mam.	40	10	1	......	10.5	2670	....	2670	5.1
19	No. 5	40	15	1	......	14.5	3640	....	3640	5.4
20	Mam.	40	15	1	......	15.0	3610	....	3610	5.0
21	No. 5	40	20	1	......	18.5	4270	....	4270	4.9
22	...do...	50	2	20	......	....	1150	....	3425	7.3
23	...do...	50	6	1	......	7.5	2320	230	2320	5.6
24	Mam.	50	6	1	......	8.0	2345	190	2345	7.4
(*b*) 25	No. 5	40	− 6	2	2	....	300	....	600	3.5
26	...do...	40	− 5	3	3	....	345	....	900	3.0
27	...do...	40	− 4	10	7	....	400	....	1735	3.8
28	...do...	40	− 3	12	14	....	480	....	2270	3.1
29	...do...	40	− 2	13	12	....	585	....	2435	3.8
30	...do...	40	− 1	13	......	2.0	735	105	2145	4.7
31	...do..	40	0	16	......	2.7	890	202	2420	4.6
32	...do...	40	1	7	......	3.5	1065	113	2165	4.5
33	...do...	40	2	9	......	4.5	1290	121	2100	4.3
34	...do...	40	3	4	......	5.0	1500	....	1805	3.9
35	...do...	40	4	1	......	5.7	1720	194	1720	4.0
36	...do...	40	5	1	......	6.5	1995	....	1995	....
37	...do...	40	5	1	......	7.0	1930	169	1930	4.5
38	...do...	40	5	1	......	6.5	1960	195	1960	3.8
39	...do...	40	6	1	......	7.7	2195	165	2195	3.8
40	...do...	40	10	1	......	11.2	2950	129	2950	3.7
41	...do...	40	10	1	......	11.0	2945	....	2945	3.6
42	...do...	40	15	1	......	15.0	3700	....	3700	3.2
43	...do...	40	15	1	......	15.2	3775	....	3775	3.1
44	...do...	40	20	1	......	19.5	4430	....	4430	2.7
45	...do...	40	20	1	......	19.5	4355	....	4355	2.6

(*a*) Battery Rogers. Height of axis of gun above water 36 feet. Shots 1 to 12 on August 28, 1865; day calm. Shots 13 to 24 on September 4, 1865; wind gentle from the front.

(*b*) Fort Foote, Maryland. Height of axis of gun above water 103 feet. Shots 25 to 29 on September 19, 1865; wind gentle from front. Shots 30 to 45 on September 14, 1865; wind gentle from rear.

Practice with the 15-inch gun.

(TABLE No. 2.)

No. of shots.	Elevation referred to horizon.	No. of impacts. 1st observer.	No. of impacts. 2d observer.	Distance in yards between measured impacts beginning at gun. 1.	2.	3.	4.	5.	6.	7.	8.	9.	10.	11.	12.	13.	14.	Last.
1	2°	19+	23	955	675	385	275	200	150	90	110	60	60	60	...	...	...	220
2	2	13+	17	1040	640	420	285	255	170	110	90	...	...	...	...	...	...	280
3	2	24	19	1100	650	310	295	180	135	95	85	60	50	30	50	35	35	120
4	2	21	16	1105	630	400	195	130	100	55	85	80	50	40	55	50	60	225
Mean...	2	...	...	1050	649	379	263	191	139	85	92	...	...	...	...	...	...	...
5	3	10	9	1340	490	295	160	90	55	35	25	20	...	...	...	...	...	20
6	3	11	10	1375	520	275	160	100	55	45	30	15	10	...	...	...	...	30
7	3	12	11	1350	510	285	175	95	65	50	30	25	15	15	...	...	...	40
Mean...	3	...	...	1355	507	285	165	95	58	43	28	20	...	...	...	...	...	...
8	4	7	6	1580	400	180	70	30	25	...	...	...	...	...	...	...	...	35
9	4	8	...	1565	375	150	70	40	15	35	...	...	...	...	...	...	...	20
Mean...	4	...	...	1572	387	165	70	35	20	...	...	...	...	...	...	...	...	...
10	5	2	2	1770	90	...	...	...	...	...	...	...	...	...	...	...	...	0
11	5	2	2	1855	35	...	...	...	...	...	...	...	...	...	...	...	...	0
12	5	2	2	1880	80	...	...	...	...	...	...	...	...	...	...	...	...	0
Mean...	5	2	2	1835	68	...	...	...	...	...	...	...	...	...	...	...	...	...
22	2	20	...	1150	700	410	260	185	125	90	60	110	35	35	...	...	...	265
25	−6	2	2	300	300	...	...	...	...	...	...	...	...	...	...	...	...	0
26	−5	3	3	345	475	80	...	...	...	...	...	...	...	...	...	...	...	0
27	−4	10	7	400	770	290	120	55	50	...	...	...	...	...	...	...	...	50
28	−3	12	14	480	910	370	215	80	...	...	...	...	...	...	...	...	...	215
29	−2	13	12	585	710	365	215	160	...	...	...	...	...	...	...	...	...	400
30	−1	13	...	735	695	345	120	80	40	20	...	...	...	...	...	...	...	110
31	0	16	...	890	700	350	195	95	60	30	35	20	...	...	...	...	...	45
32	1	7	...	1065	715	220	100	30	20	15	...	...	...	...	...	...	...	0
33	2	9	...	1290	470	190	50	45	...	...	...	...	...	...	...	...	...	55
34	3	4	...	1500	250	...	...	...	...	...	...	...	...	...	...	...	...	55

ANALYSIS OF THE PROBLEM OF REBOUND.

Before proceeding to discuss these data with a view to investigating the laws governing rebounds from water, a few general inferences respecting the problem may be drawn.

After the first impact, the trajectory must be a very complex one. The shot strikes the water, buries itself a certain distance, and then emerges, chiefly because the line of least resistance follows an upward direction. This is repeated at every impact, until the upward component of the diminished velocity is unable to overcome the action of gravity upon the shot, which accordingly sinks. If any anomalous force acts at any particular plunge, it must exert a direct and multiplied influence upon the whole of the remaining trajectory ; an influence which, with forces so delicately adjusted, must produce an important modifying effect. For instance, it is well established by observation that moderate waves may reduce more than one-half, or may nearly double the normal length of a rebound, according to the inclination of the water surface struck toward the horizon. The direction of the rotation of the shot about its axis, and the inclination of this axis, must also have a powerful effect upon the rebound. Other causes of disturbance might be named, and hence we must conclude that even on a calm day, with projectiles having the same shape, size, and density, the same striking velocity at first impact and the same angle of fall, causes beyond our control exist which prevent any two from following exactly the same path.

The first twelve shots of the last table illustrate these views. They prove that to determine the exact normal trajectory for a given elevation, charge of gun, height of battery, and initial velocity of projectile, the mean of a large number of shots must be taken. Such series of observations, being both expensive and laborious, have never been attempted, and in the present investigation we must therefore be content to deal with *approximate trajectories* only ; a necessity which still further complicates a problem already difficult.

It is probably for these reasons that the laws governing rebound from water have never received a mathematical analysis from writers upon artillery. The only paper known to me which treats of the subject in any but the most general terms is an article* in the Coast Survey Report for 1864 by Assistant C. A. Schott; and this is limited to the particular case of the probable trajectory of a ball, the lengths of whose successive bounds are known by measurement. Even this comparatively simple discussion is based on the assumption—in my judgment utterly inadmissible—that for each impact the angles of incidence and of rebound are equal.

Impressed with the importance of knowing definitely how much the advantages of ricochet fire against unarmored vessels are sacrificed in locating sea-coast batteries above the water level, with a view to gain a plunging fire upon iron-clad vessels, and possessed of the above series of experiments, (very exact, although too limited in number to exhibit precisely the normal rebounds,) I have devoted what leisure time my professional duties permitted,† to a mathematical analysis of the problem of ricochet on water. The formulæ thus framed are believed to be correct in principle; their constants are deduced from the experiments detailed above; and their claims to practical value must rest upon the accordance of their indications with actual measurement.

The first step in the analysis consisted in computing for each shot, or, when the data permitted, for the mean of each group of shots fired under identical circumstances, the elements of the trajectory for the first branch, *i. e.*: the part between the gun and the first point of impact.

---

* The same paper appears in the report of Brigadier General A. P. Howe, Inspector of artillery United States army, to the chief of engineers, as an application by him of Poissons formulæ to the case of a ricochet shot. This report is contained in Major King's compilation of recent reports on experimental firing with modern sea-coast artillery, printed for the use of officers of the Corps of Engineers.

† In the arithmetical labor I have been aided by Sergeant F. W. Lehnartz, company C, Battalion of Engineers.

For this purpose Didion's formulæ received the preference, as General Haskin's computations indicate, that, except at great elevations, they accord well with actual firing of the 15-inch gun. They are given in Didion's "Traité de Balistique," and are republished in Benton's "Ordnance and Gunnery," with convenient tables for correcting for the resistance of the air. The following are such of these formulæ as have been employed in the investigation:

$$(1). \qquad y = x \tan \varphi - \frac{gx^2}{2V^2 \cos^2 \varphi} B.$$

$$(2). \qquad \tan \theta = \tan \varphi - \frac{gx}{V^2 \cos^2 \varphi} I.$$

$$(3). \qquad t = \frac{x}{V \cos \varphi} D.$$

$$(4). \qquad v = \frac{V \cos \varphi}{\cos \vartheta U}.$$

The following is a complete list of the symbols used in these formulæ, and in the subsequent analysis; the latter added for convenience of reference. Figure 26, Plate I, has been prepared to illustrate some of these symbols to the eye. In order to designate with precision the branch under consideration, the corresponding dash—as $'$, $''$, $'''$, ... ^{n}—is placed below and to the right of quantities requiring it. Thus $V_{n'}$ $\varphi_{n'}$ $\vartheta_{n'}$ $v_{n'}$ and f_n designate the values of these quantities for the nth branch *i. e.*, the branch following the n–1 rebound. Unless stated to the contrary, the point p, named below, is considered in the computations to be a point of impact.

V = initial velocity—for any designated branch—in feet per second.

v = velocity at any point p in feet per second.

ϕ = angle at the origin of the given branch, included between the tangent to the trajectory, and a line drawn to next point of impact.

θ = angle included between the tangent to the trajectory at the point p, and a straight line connecting the origin and end of the branch considered.

ε = vertical angle included between a horizontal plane and a straight line drawn from the point of first impact to the muzzle of the gun.

x and y = rectangular co-ordinates of any point p, expressed in feet In the first branch the axis of abscissas is the straight line drawn from the muzzle of the gun to the point of first impact; in the other branches it is a straight line connecting the points of impact. The origin of co-ordinates for each branch is taken at its first point.

X and Y = maximum (but not corresponding) values of x and y in branch considered—*i. e.*, the range and extreme height—expressed in feet.
g = velocity generated by gravity in one second—assumed at 32.18 feet.
t = time of flight in seconds from origin of branch considered to any point p.
T = time of flight in seconds corresponding (in each branch) to X.
e = co-efficient of elasticity.
f = co-efficient of penetration.
R = Radius of projectile in feet.
W = Weight of projectile in pounds.
B, I, D, U = co-efficients for correcting for resistance of air.
(See tables of values in Benton's Ordnance and Gunnery.)

The initial velocity and other elements of the trajectory for each shot or group of shots were computed separately by these formulæ, allowing for the resistance of the air; and the results, so far as they bear on the succeeding analysis, are shown in the following table. It will be noticed that the times of flight closely accord with the observations given in the former table, and thus confirm the accuracy of the formulæ:

Elements of observed trajectories.

Number of shot.	Elevation above horizon, $\phi_{,}-\varepsilon$.	Height of battery.	Range to first impact, $X_{,}$.	Initial velocity, $V_{,}$.	Angle of fall. On plane passing through gun and point of impact, $=\theta_{,}$.	Angle of fall. On water, $\theta_{,}+\varepsilon$.	Velocity at 1st impact. $v_{,}$.	Velocity at 1st impact. $v_{,}\cos\theta_{,}$.	Velocity at 1st impact. $v_{,}\sin\theta_{,}$.	Time of flight, $T_{,}$.
	°	*ft.*	*ft.*	*ft.*	° ′	° ′	*ft.*	*ft.*	*ft.*	*sec.*
1–2–3–4=1st mean	2	36	3150	1171	3 16	3 55	859	857	58	3. 2
5–6–7=2d mean	3	36	4065	1199	4 34	5 04	807	804	71	4. 2
8–9=3d mean	4	36	4716	1176	5 59	6 25	750	745	84	5. 1
10–11–12=4th mean	5	36	5505	1191	7 33	7 55	708	702	98	6. 1
13–15=5th mean	6	36	6186	1196	8 28	8 48	671	662	103	7. 1
14–16=6th mean	6	36	5886	1146	9 04	9 25	667	658	109	6, 9
22d shot	2	36	3450	1250	3 16	3 52	888	886	60	3. 3
25th shot	— 6	104	900	1216	0 39	7 15	1106	1103	140	0. 8
26th shot	— 5	104	1035	1175	0 48	6 32	1054	1051	120	0. 9
27th shot	— 4	104	1200	1122	1 01	5 58	995	989	103	1. 1
28th shot	— 3	104	1440	1138	1 15	5 23	985	981	92	1. 4
29th shot	— 2	104	1755	1146	1 34	4 58	961	957	83	1. 7
30th shot	— 1	104	2205	1184	1 58	4 40	947	943	77	2. 1
31st shot	0	104	2670	1130	2 46	5 00	863	859	75	2. 7
32d shot	1	104	3195	1135	3 32	5 24	832	828	78	3. 3
33d shot	2	104	3870	1160	4 27	5 59	798	793	83	4. 1
34th shot	3	104	4500	1152	5 38	6 57	750	744	91	4. 9
35th shot	4	103	5160	1165	7 08	8 16	693	685	100	5. 8
36–37–38=7th mean	5	103	5886	1179	8 40	9 40	681	671	114	6. 7

Certain general deductions follow from an inspection of this and of the preceding tables of data. Thus, the fourth mean and the thirty-fourth shot have ricochets of sixty-eight and two hundred and fifty yards; and the fifth mean and thirty-fifth shot both fail to rebound. The angles of fall upon the water surface ($\vartheta_{,}+\varepsilon$) are in the former 7° 55′ and 6° 57′, and in the latter 8° 48′ and 8° 16′. This confirms what the text-books state, (based upon experiments with smaller guns,) that rebounds cease when the angle of fall is about 8°;—and that, too, whether the shot is fired from thirty-six or one hundred and four feet above the water surface. In other words, ricochet depends upon the angle of fall and velocity of impact of the projectile, whether the gun is placed on a high or on a low site; a law naturally to be expected. Again, assuming the fourth mean as marking very nearly the limit of ricochet, we see that, to allow a rebound, the vertical component of the velocity of impact ($v_{,}$ sin $\vartheta_{,}$) must not exceed about one-seventh of the horizontal component, ($v_{,}$ cos $\vartheta_{,}$) Applying this rule to the twenty-fifth shot, where $v_{,}$ cos $\vartheta_{,}$ equals eleven hundred feet, $v_{,}$ sin $\vartheta_{,}$ must not exceed one hundred and sixty feet. It was actually one hundred and forty feet, and the shell rebounded three hundred yards. Other useful conclusions not necessary to mention may be drawn from this table, which exhibits considerable range in respect to angles of incidence and relative horizontal and vertical velocity.

The proposed general analysis really begins at this point, and is based upon the following principles:

I. The velocity has already undergone so great a reduction before the first impact, and the loss of velocity during this impact must be so considerable, that it is allowable to neglect the resistance of the air as a separate co-efficient in tracing the remainder of the trajectory,—especially considering the necessary uncertainties involving it. Since the numerical values of the constants of the new formulæ are deduced from the observed bounds, &c., it will be noticed that they correct for the effect of a resistance of the air due to the actual velocities of emergence, (V_n.) Variations in this velocity are practically confined to so narrow limits that this manner of correcting for it is sufficiently exact, and far less laborious than the introduction of the correction in the usual form; which might, of course, be done if desired. Formulæ (1), (2), (3), and (4), become applicable by omitting the co-efficients B, I, D and U. Thus simplified, they will be designated (1′), (2′), (3′), (4′).

II. The problem bears a certain analogy to that of a hard ball impinging upon a large slab of the same material, a particular case of impact of solids, for which we have well-established formulæ, viz: (See list of symbols already given, remembering that $\vartheta_{,}+\varepsilon$ must be used in place of $\vartheta_{,}$ for first impact.)

$$(5). \qquad V_n = v_{n-1}\sqrt{e^2_n \sin^2 \vartheta_{n-1} + \cos^2 \vartheta_{n-1}}.$$

$$(6). \qquad \text{Cot.}\, \varphi_n = -\frac{\cot \vartheta_{n-1}}{e_n}.$$

It is to be remarked, however, that these formulæ are deduced upon the supposition that there is no *indentation* of the slab, and hence that there is no loss in the component of velocity parallel to its surface. A shot impinging upon water buries itself in the fluid, and hence experiences a great resistance to horizontal motion. A coefficient of $\cos^2\vartheta_{n-1}$ must therefore be introduced to adapt *eq.* (5) to the case in question; that is, it must become:

$$(5'). \qquad V_n = v_{n-1}\sqrt{e^2_n \sin^2 \vartheta_{n-1} + f_n^2 \cos^2 \vartheta_{n-1}}.$$

For convenience of computation *eq.* (5′) is put under the form:

$$(5''). \qquad V_n = f_n\, v_{n-1} \cos \vartheta_{n-1}\sqrt{\frac{e^2_n}{f_n^2} \tan^2 \vartheta_{n-1} + 1}.$$

If the proper values of e_n and f_n in *eq.* (5″) and (6) can be discovered, the problem is solved; for these formulæ, in connection with (1′), (2′), (3′), and (4′), will enable us to fully discuss all the branches of the trajectory.

In the impact of solids, e is the modulus of elasticity of the material, and its numerical value may be easily deduced by a simple experiment. It is constant for any particular material.

In the case of a solid impinging upon a liquid, the circumstances are quite different. The value of e_n depends on the conditions—other than velocity and angle of incidence—which influence the direction of the line of least resistance through the water; that is, it is a function of the elasticity, form and density of the projectile, and of the density and elasticity of the fluid. Hence, for iron balls of the same weight and calibre impinging upon water, the inference is legitimate that e_n has a sensibly constant numerical value, but one which cannot, as in the case of solids, be directly deduced by experiment. We may, however, reasonably infer that this value must exceed unity, because, otherwise *eq.* (6) would exact that the angle of rebound should be less than that of incidence; whereas all authorities agree that the reverse must be the case, owing to the reduction of velocity during the plunge; an opinion which the mark left by a shot ricocheting upon soft ground confirms.

To deduce the numerical value of e_n, therefore, we must depend upon an analysis of the data in hand, which was done by two distinct processes. The first depends upon the condition that e must have such a value in *eq.* (6) as to make the number of successive values of φ for each shot, between its first value and the limit of ricochet, accord with the actual number of observed impacts. The actual number of ricochets for some of the shots was only approximately known, owing to the difficulty of accurately counting those near the point of sinking; but a careful discussion of the data indicated that, to secure this accordance, e could not differ much from 1.05.

The second process for deducing the value of e_n is purely algebraic. Making, in *eq.* (1′), $y=o$ and reducing, it becomes:

$$(7). \qquad X_n = \frac{\sin 2\varphi_n V_n^2}{g}.$$

Combining this equation with *eq.* (5′) and *eq.* (6), eliminating V_n and φ_n, and neglecting certain small differences in order to avoid an equation of the third degree, the following approximate value of e_n results, in which, when applied to the data of the last table, f^2_n is the only unknown quantity.

$$(8).\ e_n = \left\{ \frac{gX_n}{2 \sin 2\vartheta_{n-1} (\sin \vartheta_{n-1} v_{n-1})^2} - \sqrt{\left(\frac{gX_n}{2 \sin 2\vartheta_{n-1} (\sin \vartheta_{n-1} v_{n-1})^2} \right)^2 - f_n^2 \cot \vartheta_{n-1}} \right.$$

As will soon be explained, it proved to be necessary to deduce several successive values for f_n^2, each approximating more nearly to its true value. Using these values successively in this equation, after much tedious arithmetical work, it was found that the final values of e_n, which differ but slightly among themselves for the different shots, were almost the same as that indicated by the first process, viz., 1.05, which was accordingly adopted with every confidence in its correctness.

The value of f_n^2 involved much study and calculation. It is clear, since this quantity is the modulus (inversely) of the retardation of velocity due to penetration, that its value for any impact must depend upon the length of the water trajectory ; which, for a given density of ball, varies with the angle of incidence (ϑ_n). For the first impact, its value (depending upon $\vartheta_{,}+\varepsilon$) must be zero at the limit of ricochet, and nearly unity when the shot is moving essentially parallel to the water surface. After the first impact, the successive values of f^2 must increase, as the limit of ricochet is approached, until it becomes nearly unity. These conditions are accordant both with the probable action of the forces developed at the impacts and with the gradual approach to equality in the length of the successive bounds, as shown by the measurements. Hence f_n^2 cannot be a constant, but must be a complicated variable, whose successive values are represented by the ordinates of a line passing through unity at the limit of ricochet, and through a point (determined by the angle of incidence of the given shot) of an unknown curve, which passes through zero at the limit of ricochet, and near unity for an essentially horizontal impact.

To deduce the numerical equation based upon these views, which should represent f_{n}^2 correctly, was not easy, because its algebraic value involved e, a quantity at first unknown A system of approximation was adopted. The first approximate values of $f_{\prime\prime}^2$ were computed from the observations by using the following approximate equation, which results from combination, reduction and neglect of small differences in equations (5′) and (7):

$$(9). \qquad f_{\prime\prime}^2 = \frac{X_{\prime\prime} g}{\sin 2(\vartheta_\prime + \varepsilon)[v_\prime \cos(\vartheta_\prime + \varepsilon)]^2}$$

These values, computed for the first impacts, should make known the curve whose abscissas are the values of the cos $(\vartheta_\prime + \varepsilon)$, and whose ordinates are the values of $f_{\prime\prime}^2$. To represent these numerical values correctly, the equation of a right line, ($f_{\prime\prime}^2$ and not $f_{\prime\prime}$ being considered one of the variables) was assumed, and the result was favorable, although the law of variation was evidently not exactly represented by it. This equation was:

$$(10). \qquad f_{\prime\prime}^2 = 80 \cos(\vartheta_\prime + \varepsilon) - 79.2.$$

Substituting its indications for $f_{\prime\prime}^2$ in *eq.* (8), the corresponding values of $e_{\prime\prime}$, and those of $\varphi_{\prime\prime}$ in *eq.* (6), were deduced for all *first impacts*. The values of $f_{\prime\prime}^2$, given by *eq.* (9), were next multiplied by the ratio $\frac{\sin 2(\vartheta_\prime + \varepsilon)}{\sin 2 \varphi_{\prime\prime}}$ in order to correct for an assumed equality between $\varphi_{\prime\prime}$ and $\vartheta_\prime + \varepsilon$ in deducing that approximate equation. These new values of $f_{\prime\prime}^2$ necessitated new constants in *eq.* (10). They were deduced, and the operation was repeated until, as already stated, it became evident that e might, without error, be considered 1.05. The corresponding value of $f_{\prime\prime}^2$ was:

$$(10'). \qquad f_{\prime\prime}^2 = 78.08 \cos(\vartheta_\prime + e) - 77.33.$$

An equation for f_{n}^2 was then deduced subject to the condition that it should represent a right line passing through unity at the limit of ricochet (8° 6′ 30″) and through the point upon the line represented by *eq.* (10′), corresponding-to the value of $\cos \vartheta_{,}+\varepsilon$ for any shot in question. Its equation was :

$$(11).\qquad f_{n}^{2}=f_{,,}^{2}+\frac{1-f_{,,}^{2}}{0.99-\cos(\vartheta_{,}+\varepsilon)}[\cos\vartheta_{n-1}-\cos(\vartheta_{,}+\varepsilon)].$$

Using *eq.* (10′), (11) and (7), a computation was then made of all the values of X_n for all the shots ; deducing incidentally the corresponding values of f_n^2. The corresponding true values of f_n^2 were then deduced—assuming the observations to be exact and *e* to be 1.05—by applying the correction ratio represented in the following formula, in which capitals denote the true, and small type the already computed values :

$$F_n^{2}=\frac{f_{,,}^{2}f_{,,,}^{2}\,.\,.\,.\,.\,.\,.\,f^{2}_{n-1}X_n}{F_{,,}^{2}F_{,,,}^{2}\,.\,.\,.\,.\,.\,.\,F^{2}_{n-1}x_n}f_n^{2}.$$

A careful study was then made of the resulting values of F_n^2, which must be closely represented by a really correct equation for that quantity. It was found that *eq.* (11) might be adopted, but that the right line represented by *eq.* (10′) was not sufficiently exact. After a study of this curve—involving some re-computation of X_n with new trial values for $f_{,,}^2$—it was found to be a portion of an hyperbola whose axis was the ordinate at the limit of ricochets ($\cos \vartheta_n = 0.99$), and whose vertex was on the axis of abscissas. Its equation was :

$$(12).\qquad f_{,,}^{2}=\sqrt{0.1153+13260\,[\cos(\vartheta_{,}+\varepsilon)-0.99]^{2}}-0.3395.$$

Having thus determined the values of *e* and f_n^2, all the data collected were used to test the exactness of the new formulæ as applied to the 15-inch shell. The result is shown in the two following tables corresponding to the two given heights of battery, (thirty-six feet, and one hundred and four feet.) The general accordance between the formulæ and the observations exceeded my most sanguine expectations where so few *mean* trajectories were available. As already stated, single shots are always more or less affected by abnormal causes, such as waves, rotation, &c., and hence often differ considerably from the mean of a large number of shots fired under identical circumstances; which, of course, is what is represented by a correct general formula. This is well illustrated by shots 29, 30, and 31. Their elevations referred to the horizon were, respectively, —2°—1°, and 0°; their angles of fall, 4° 58′, 4° 40′, and 5° 00′; their velocities of impact, 961, 947, and 863 feet. Clearly, shot 30 should have the *longest* ricochet trajectory. It has the *shortest*, doubtless because its normal range was reduced by a rapid rotation about a horizontal axis perpendicular to the trajectory. The discrepancy between the formulæ and the observed trajectory in this case is unquestionably a merit in the former, which more nearly represents the *mean trajectory* than the latter.

Aside from shot 30, there are no discrepancies between the indications of the formulæ and the observations, which are large enough to merit attention, except in shots 25 and 26. In both of these the formulæ indicate more rebounds and a larger ricochet trajectory than the observations. It is to be remarked that the angles of fall are very large—being 7° 15′ and 6° 32′ respectively—angles which border closely on the limit of ricochet, where any abnormal cause would be very liable to abruptly shorten the trajectory. The note-books show that a "slight wind" was blowing at the time when the observations were made, and it is believed that they were affected by it, especially as the first rebounds accord well with the formulæ.

The firing was purposely conducted so as to cover, as far as practicable, the entire range of angles at which shot would ricochet, both from the gun at thirty-six feet, and from that at one hundred and four feet above the water. This wide range of angles gives the formulæ resulting from the discussion a general character, so far as the 15-inch shell is concerned; and the close accordance between their indications and the observations, makes the probability strong that the few discrepancies are due rather to the observations themselves differing from the corresponding normal trajectories, than to errors of the formulæ. In confirmation of this opinion it will be noticed that the means give the best accordances.

Comparison between measured and computed ricochet trajectories. Battery Rogers, 36 feet above water.

Branches of trajectory.	1st shot, (mean of 4.)		2d shot, (mean of 3.)		3d shot, (mean of 2.)		4th shot, (mean of 3.)		22d shot.	
	Measured bounds.	Computed. bounds.	Measured bounds.	Computed bounds.	Measured bounds.	Computed bounds.	Measured bounds.	Computed bounds.	Measured bounds.	Computed bounds.
	Yds.	*Yds.*	*Yds.*	*Yds.*	*Yds.*	*Yds.*	*Yds.*	*Yds.*	*Yds.*	*Yds.*
Number 1, (from gun) ..	1050	......	1355	......	1572	......	1835	......	1150	......
Number 2	649	665	507	554	387	299	68	38	700	711
Number 3	379	433	285	279	165	111	......	......	410	467
Number 4	263	288	165	153	70	58	......	......	260	313
Number 5	191	197	95	91	35	41	......	......	185	214
Number 6	139	137	58	59	20	36	......	......	125	150
Number 7	85	99	43	41	Lost.	......	......	......	90	108
Number 8	92	73	28	31	......	......	......	......	60	80
Number 9	Lost.	56	20	26	......	......	......	......	110	61
Number 10	. do ..	44	Lost.	24	......	......	......	......	35	48
Number 11	. do ..	36	. do ..	23	......	......	......	......	35	39
Number 12	. do ..	30	......	......	......	......	......	......	Lost.	33
Number 13	. do ..	27	......	......	......	......	......	......	. do ..	29
Number 14	. do ..	25	......	......	......	......	......	......	. do ..	26
Number 15	. do ..	24	......	......	......	......	......	......	. do ..	25
Number 16	. do ..	24	......	......	......	......	......	......	. do ..	25
Number 17	. do ..	......	......	......	......	......	......	......	. do ..	25
Total after first impact...	2, 205	2, 158	1, 259	1, 281	723	545	68	38	2, 275	2, 354
Total number of rebounds	18	15	10	10	6	5	1	1	19	16

Comparison between measured and computed ricochet trajectories Fort Foote, 104 feet above water.

Branches of trajectory.	25th shot.		26th shot.		27th shot.		28th shot.		29th shot.	
	Measured bounds.	Computed bounds.	Measured bounds.	Computed bounds.	Measured bounds.	Computed bounds.	Measured bounds.	Computed bounds.	Measured bounds.	Computed bounds.
	Yds.	*Yds.*	*Yds.*	*Yds.*	*Yds.*	*Yds.*	*Yds.*	*Yds.*	*Yds.*	*Yds.*
Number 1, (from gun)...	300	......	345	......	400	......	480	......	585	......
Number 2	300	286	475	547	770	659	910	777	710	805
Number 3	......	133	80	204	290	265	370	362	365	420
Number 4	......	118	......	111	120	133	215	189	215	235
Number 5	......	......	......	82	55	81	80	110	160	142
Number 6	......	......	......	77	50	59	Lost.	71	Lost.	92
Number 7	......	......	......	......	Lost.	52	do ..	51	. do ..	64
Number 8	......	......	......	......	. do ..	52	. do ..	41	. do ..	48
Number 9	......	......	......	......	......	......	do ..	38	. do ..	39
Number 10	......	......	......	......	......	......	. do ..	36	. do ..	35
Number 11	......	......	......	......	......	......	. do ..	......	. do ..	33
Total after first impact...	300	537	555	1021	1335	1301	1790	1675	1850	1913
Total number of rebounds	1	3	2	5	7	7	12	9	12	10

Comparison—Continued.

Branches of trajectory.	30th shot.		31st shot.		32d shot.		33d shot.		34th shot.	
	Measured bounds.	Computed bounds.	Measured bounds.	Computed bounds.	Measured bounds.	Computed bounds.	Measured bounds.	Computed bounds.	Measured bounds.	Computed bounds.
	Yds.	*Yds.*	*Yds.*	*Yds.*	*Yds.*	*Yds.*	*Yds.*	*Yds.*	*Yds.*	*Yds.*
Number 1, (from gun)...	735	......	890	......	1065	......	1290	......	1500	......
Number 2	695	799	700	640	715	552	470	423	250	193
Number 3	345	443	350	329	220	256	190	169	Lost.	77
Number 4	120	260	195	182	100	133	50	85	. do ..	53
Number 5	80	161	95	109	30	78	45	52	......	52
Number 6	40	105	60	70	20	51	Lost.	38	......	......
Number 7	20	73	30	49	15	38	. do ..	33	......	......
Number 8	Lost.	54	35	37	......	31	......	......	......	......
Number 9	. do ..	42	20	30	......	28	......	......	......	......
Number 10	. do ..	35	Lost.	27	......	......	......	......	......	......
Number 11	. do ..	31	. do ..	26	......	......	......	......	......	......
Number 12	. do ..	30	. do ..	......	......	......	......	......	......	......
Total after first impact...	1410	2033	1530	1499	1100	1167	810	800	305	375
Total number of rebounds	12	11	15 (?)	10	6	8	8 (?)	6	3	4

It is claimed that these results are sufficient to establish the accuracy of the new formulæ for the case of the 15-inch shell, whatever be the height of the battery or the charge of the gun. Their application to other projectiles can of course be determined only by experiment. To test the matter, General Haskin, commanding fort Schuyler, kindly offered to fire for me, provided the necessary authority to expend the ammunition were obtained. The Chief of Ordnance gave this authority, and a series of experiments was accordingly conducted with solid shot from the 8-inch columbiad and the 24-pounder siege gun. All the precautions (already described) observed in obtaining the data for the 15-inch shell were used, and the results may be relied upon as exact. To eliminate the abnormal effect of rotation, the several shots were fired under identical circumstances at a single selected angle of elevation; thus giving one well-determined mean trajectory for each calibre. The time of flight was also noted as closely as possible with a common watch. The 8-inch columbiad was fired from a height above the water of 38.5 feet, with an angle of elevation respecting the horizon of 0°, a charge of ten pounds of cannon powder No. 5, (Oriental Powder Company, 1862,) and a solid shot weighing sixty-four pounds. The 24-pounder siege gun was fired from a height of 15.2 feet above the water, with an angle of elevation respecting the horizon of 2°, a charge of six pounds of above-named powder, and a solid shot weighing 24.5 pounds. A single 24-pounder shot (marked No. 19 in table) was fired at an elevation of 4°, the other conditions remaining the same. The 8-inch columbiad was fired on August 24, 1867, and the 24-pounder on August 26, 1867; both very calm days with smooth water. The following tables exhibit the data collected.

A fine example of the effect of rotation about a vertical axis was given by shots 16 and 17. The planes of fire(from gun to first impact) were practically identical. Shot 16 twisted to the left, and shot 17 to the right, until at their points of sinking the perpendicular distance from the plane of fire, was for each about fifty yards; thus separating laterally one hundred yards from each other in a distance of about four hundred and fifty yards. It is easy to understand what a difference in ricochet trajectory would have resulted had the axis of rotation been horizontal instead of vertical.

Practice with the 8-inch Columbiad.

No. of shot.	No. of impacts.		Distance in yards and time of flight in seconds between measured impacts, beginning at gun.															Total range and time of flight.
	1st observer.	2d observer.	1.	2.	3.	4.	5.	6.	7.	8.	9.	10.	11.	12.	13.	14.	Last.	
1	13	11	580	530	420	250	200	120	120	80	80	30	30	30	......	......	0	2470 15
2	18	21	560 2.5	650 1.5	430	280	110	70	50	40	30	40	50	40	......	......	480	2830 17
3	19	19	610 1.5	740 1.0	340	240	190	80	30	40	60	60	40	......	......	......	150	2580 11
4	18	18	560 2.0	510 2.0	270	250	190	220	130	120	130	100	50	80	60	50	80	2800 17
5	13	13	610 1.5	400 1.0	330	330	270	290	180	110	80	90	40	......	......	......	60	2790 16
6	24	23	580 1.5	450 1.0	320	260	250	210	150	90	110	120	......	......	......	......	370	2910 19
7	22	23	600 2.0	560 1.5	310	280	200	180	130	60	50	80	70	80	70	70	180	2920 18
8	11	10	500 2.0	600 1.5	340	240	260	180	120	20	30?	20	......	......	......	......	0	2310 13
9	12	13	600 2.0	430	470	190	170	200?	180	170	140	80	50	......	......	......	110	2790 16
Mean...	17	17	578 1.7	541 1.4	359	258	204	172	121	81	79	69	......	......	......	......	......	2711 15.5

Practice with the 24-pounder gun, elevation 2°.

No. of shot.	No. of impacts.		Distance and time of flight in seconds between measured impacts, beginning at gun.															Total range and time of flight.
	1st observer.	2d observer.	1.	2.	3.	4.	5.	6.	7.	8.	9.	10.	11.	12.	13.	14.	Last.	
10	11	11	1090 4.0	440 2.0	280	160	120	60	40	20?	......	......	......	......	......	......	40	2250 16
11	15	15	1130 3.0	410 2.0	280	150	120	100	90	70	......	......	......	......	......	......	130	2480 15
12	17	16	1220 2.5	500 1.5	220	100	60	50	50	50	......	......	......	......	......	......	150	2400 14
13	12	13	1110 3.0	500 1.5	300	210	90	50	40	40	20	......	......	......	......	......	90	2450 13
14	13	14	1190 3.0	460 2.0	230	130	170	50	40	20	30	......	......	......	......	......	60	2380 15
15	10	12	1190 3.0	380 2.0	220	120	60	50	20	20?	......	......	......	......	......	......	10	2070 14
16	16	13	1090 2.5	510 1.5	230	180	140	110	50	50	50	30	......	......	......	......	130	2570 14
17	10	12	1240 3.0	420 2.0	250	150	80	70	50	20	......	......	......	......	......	......	40	2320 13
18	13	14	1060 3.0	500 1.5	280	160	130	90	50	50	30	10	......	......	......	......	50	2410 14
Mean...	13	13	1147 3.0	458 1.8	254	151	108	70	48	38	......	......	......	......	......	......	......	2370 14.2

24-pounder siege gun, elevation 4°.

No. of shot.	1st observer.	2d observer.	1.	2.	3.	4.	5.	6.	7.	8.	9.	10.	11.	12.	13.	14.	Last.	Total range and time of flight.
19	2	2	1640	50	......	......	......	......	......	......	......	......	......	......	......	......	0	1690

The formulæ (1), (2), (3), and (4) were applied to these data with the following results:

Elements of observed trajectories.

Number of shot.	Elevation above horizon, $\phi_{\prime}-\varepsilon$	Height of battery.	Range to 1st impact, $X_{\prime}$	Initial velocity, $V_{\prime}$	Angle of fall. On plane passing through gun and point of impact, $\theta_{\prime}$	Angle of fall. On water, $\theta_{\prime}+\varepsilon$	Velocity at 1st impact, $v_{\prime}$	Time of flight, $T_{\prime}$
	° ′	*Ft.*	*Ft.*	*Ft.*	° ′	° ′	*Ft.*	*Sec.*
8-inch shot, (mean of 9).	0 0	38.5	1734	1233	1 32	2 48	945	1.6
24-pdr. shot, (mean of 9).	2 0	15.2	3441	1592	3 39	3 54	771	3.2
24-pdr. shot, (single)	4 0	15.2	4920	1592	7 53	8 04	599	5.4

In applying the new formulæ to these data it is clear that if of general application, no modification in them can be required, except it be a legitimate change in the values of e and f resulting from the difference between the projectiles; *i. e.*, the 15-inch shell and the 8-inch and 24-pounder solid shot. These projectiles differ in size and in density. It is to be remembered that e and f are not moduli of the resistances encountered by the projectile, but are coefficients depending upon its line of direction and retardation of velocity in passing through the water. The resistance to penetration is proportional to the area of the surface striking the water; the power to overcome it is proportional to the mass of the projectile. The resistance and the power, therefore, vary as R^2 and R^3 respectively, for spherical shot of equal density. Hence the measure of the ability of such a projectile to overcome the resistance of the water is a function of $\frac{R^3}{R^2}$. This ratio is for the 15-inch solid shot and the 24-pounder solid shot, respectively, 0.62 and 0.24. But the effect of this greater power in the 15-inch solid shot to overcome the resistance reduces the primary deflection, lengthens the water trajectory, and thus tends to equalize the retarding effect of the water upon the velocity of the two projectiles. In other words, there is a certain compensation which indicates that e and f probably remain nearly constant for shots of equal density and different sizes.

If, however, an increase in density occurs, as is the case when a solid shot is substituted for a shell of the same calibre, the water trajectory is lengthened without a corresponding reduction in surface, the velocity of emergence is proportionally reduced, and the coefficients e and f must vary accordingly. What will be the effect upon e—that is, how the density of the projectile enters its equation—can hardly be predicted, except that its value must be increased. The effect upon $f_{\prime\prime}^2$, however, would evidently be in the ratio of $\frac{D'}{D}$, in which D′ denotes the density of the 15-inch shell, and D that of the corresponding solid shot. The density of a sphere being given by the equation

$$D=\frac{W}{\frac{4}{3}\pi g R^3},$$

and the weights of the loaded shell used in the experiments, and the solid shot being 344 and 425 pounds respectively, the ratio $\frac{D'}{D}$ becomes 0.81. For a solid shot *eq.* (12) therefore takes the form,

$$(12'). \quad f_{\prime\prime}^2=0.81\left(\sqrt{0.1153+13260\left[\cos(\vartheta_{\prime}+\varepsilon)-0.99\right]^2}-0.3395\right)$$

The algebraic values of f after the first impact, given by *eq.* (11) undergo no change, because the reduction in the numerical value of $f_{\prime\prime}^2$, which enters this equation renders it applicable to the solid shot.

In fine, then, we should anticipate that for any solid shot the formulæ for the 15-inch shell must remain unchanged, except that *eq.* (12) becomes *eq.* (12′), and that the numerical value of e in *eq.* (5″) and *eq.* (6) increases, but probably not to any large extent.

The following tables exhibit the results of the computation for the Fort Schuyler observations:

Comparison between measured and computed ricochet trajectories, Fort Schuyler, N. Y.

Branches of trajectory.	8-inch columbiad. (Mean of 9 shots.)			24-pdr. gun. Elevation 2°. (Mean of 9 shots.)			24-pdr. gun. Elevat'n 4°. (Single shot.)		
	Measured bounds.	Computed bounds. $e = 1.07$.	Computed bounds. $e = 1.05$.	Measured bounds.	Computed bounds. $e = 1.07$.	Computed bounds. $e = 1.05$.	Measured bounds.	Computed bounds. $e = 1.07$.	Computed bounds. $e = 1.05$.
	Yds.	*Yds.*	*Yds.*	*Yds.*	*Yds.*	*Yds.*	*Yds.*	*Yds.*	*Yds.*
No. 1 from gun	578			1147			1640		
2	541	573	563	458	442	435	50	25	24
3	359	369	355	254	244	233			
4	258	241	225	151	141	129			
5	204	160	145	108	86	74			
6	172	108	94	70	56	44			
7	121	75	62	48	38	27			
8	81	53	42	38	28	18			
9	79	39	29	Lost	22	12			
10	69	29	20	do	19	9			
11	Lost	23	14	do	17	6			
12	do	18	10	do	17	5			
13	do	15	7	do	17	4			
14	do	13	6			4			
15	do	12	4			3			
16	do	12	3			3			
17	do	12	3						
18			2						
19			2						
20			2						
21			2						
22			2						
Total after 1st impact	2133	1752	1592	1223	1127	1006	50	25	24
Total number of rebounds	16	16	21	12	12	15	1	1	1

The following are the times of flight as observed and computed, using *eq.*(3) for the 1st branch and *eq.*(3′) for the remaining branches. It is to be remembered that as these times were noted with a common watch, the observations are only approximate.

Times of flight.

Gun.	1st branch.		2d branch.		Total.	
	Observed.	Computed.	Observed.	Computed.	Observed.	Computed.
	Sec.	*Sec.*	*Sec.*	*Sec.*	*Sec.*	*Sec.*
8-inch columbiad, (mean of 9 shots)	1.7	1.6	1.4	2.4	15.5	17.5
24-pounder gun, elevation 2°, (mean of 9 shots.)	3.0	3.2	1.8	2.4	14.2	16.2

These results certainly show a gratifying accordance between the measurements and the indications of the new formulæ. For several of the first impacts the agreement may be considered perfect. Toward the end the formulæ indicate rather too short bounds, thus unduly reducing the total range. It is to be remarked, however, that the observations differ greatly among themselves in this quantity, which seems to be less uniform with these calibres than with the 15-inch shell. The latter part of the trajectory has no value against shipping.

The table shows that the accordance is decidedly improved, both in range and number of impacts, by making $e=1.07$, instead of 1.05. A still larger value unduly reduces the number of impacts. This variation, then, represents the practical effect upon that quantity of the difference in density between the 15-inch shell and the solid shot. The change is evidently too slight to call for the framing of an equation. It is sufficient to consider e equal to 1.05 for shells and 1.07 for solid shot.

The table proves that, as surmised, the value of $f_{//}^2$ corresponding to the 15-inch shell, becomes affected by a coefficient $\frac{D'}{D}=0.81$, when a solid shot is used. This coefficient applies to solid shot of all calibres. For shells having the same density as the 15-inch shell, the coefficient remains unity. For other shells it must have the following value, in which $W_/$ and $R_/$ denote respectively the weight in pounds and the radius in feet of the 15-inch shell used in my firing, and $W_{//}$ and $R_{//}$ the same quantities for the given shell.

$$\frac{\dfrac{W'}{\frac{4}{3}\pi g R_/^3}}{\dfrac{W_{//}}{\frac{4}{3}\pi g R_{//}^3}}=1446\frac{R_{//}^3}{W_{//}}.$$

APPLICATION OF THE NEW FORMULÆ.

The new formulæ are of practical utility in the solution of some of the most perplexing problems now offered to the

Corps of Engineers for solution. Thus, for instance, it has heretofore been generally admitted that the *lower* channel-bearing guns can be placed—provided the cannoniers are protected against a plunging fire from the tops of an enemy's fleet—the more efficient will they be for defence ; and most of our existing water batteries have been constructed accordingly. This rule has received important modifications in the recent introduction of armored vessels capable of taxing the powers of the heaviest of modern guns—as, for instance, the Kalamazoo class of monitors, and the new patterns of broadside ships of war with heavily plated and greatly inclined iron casemates. Ricochet fire is of no avail against them. With any but our very heaviest guns the deck or the casemate itself must be directly struck, and that too by a shot retaining a high velocity and moving on a line as nearly as possible normal to its surface. In other words, we must use a *plunging fire*, which necessitates a certain height to the battery. For economy in defence, also, much is gained by raising the guns above the water level. A low battery is directly exposed to a concentrated fire from heavy ordnance which cannot be resisted by stone walls, and which it seems probable must be opposed by iron, either as plating to the masonry or as a substitute for it. Place the battery one hundred or one hundred and fifty feet above the water and all this is changed. It can only be assailed by the fleet under such disadvantages that shots striking it will be reduced in velocity beyond the dangerous limit, so that little or no plating will be necessary. We thus, by giving them a certain elevation, increase the efficiency and reduce the cost of water batteries so far as iron-clad vessels are concerned. Against unarmored vessels ricochet fire still remains the best that can be used, owing to the greater chances of hitting the object ; and it is therefore of importance to know definitely how much of its efficiency is sacrificed by raising the guns above the water. This definite knowledge, heretofore unattainable, is given by the new formulæ.

As an illustration of their uses a solution to the following problem will be given. Suppose a 15-inch gun to be fired with forty pounds of No. 5 powder, and a loaded shell weighing three hundred and forty-four pounds—required the elements of its longest ricochet trajectory at each of a series of heights varying from the water level to the greatest elevation at which a rebound can be obtained.

The first step is to ascertain the most favorable angle for ricochet firing at the different heights selected, and the maximum height at which a rebound can be obtained. This can be done in the following manner.

For the general case, use should be made of formulæ (1) and (2) to compute the horizontal range and angle of fall of the given projectile for every thirty minutes of elevation between 0° and 6°; but with the present problem this is not necessary, as the computations have already been made, based upon the actually observed ranges. The results are shown in the following table, which, it will be noticed, is a carefully deduced table of experimental ranges of the 15-inch gun.

Data respecting trajectory of 15-*inch gun, derived from observations already detailed and referred to plane passing through muzzle of gun and point of impact.*

Number of shot. (See table I of data.)	Elevation above plane of reference. $\phi_{\prime}$	Angle of fall upon plane of reference. $\theta_{\prime}$	Range to 1st impact. $X_{\prime}$	Striking velocity. $v_{\prime}$	Initial velocity. $V_{\prime}$
	° ′	° ′	*Feet.*	*Feet.*	*Feet.*
25	0 36	0 39	900	1106	1216
26	0 44	0 48	1035	1058	1175
27	0 57	1 01	1200	994	1122
28	1 08	1 15	1440	985	1138
29	1 24	1 34	1755	960	1146
30	1 42	1 58	2205	946	1184
31	2 14	2 46	2670	862	1130
1-2-3-4=1st mean..	2 39	3 16	3150	858	1171
32	2 52	3 32	3195	832	1135
5-6-7=2d mean.....	3 30	4 34	4065	807	1199
33	3 32	4 27	3870	798	1160
34	4 19	5 38	4500	750	1152
8-9=3d mean.......	4 26	5 59	4716	750	1176
35	5 08	7 08	5160	685	1165
10-11-12=4th mean.	5 22	7 33	5505	708	1191
36-37-38=7th mean.	6 00	8 40	5886	681	1146
13-15=5th mean....	6 20	8 28	6186	670	1196

It is easily demonstrated that the most favorable angle for ricochet firing for any given height of battery and usual initial velocity, is that which gives its minimum value to the angle of fall upon the water surface. For any trajectory this angle of fall upon the water is represented by $\vartheta_{,}+\varepsilon$, and the corresponding angle of elevation of gun referred to the horizon by $\varphi_{,}-\varepsilon$. Computing the values of ε from the above values of $X_{,}$ and the known height of the battery, and taking the corresponding values of $\varphi_{,}$ and $\vartheta_{,}$ from the table, we can, by simple addition and substraction, deduce the value of $\varphi_{,}-\varepsilon$ that answers to each, and hence to the minimum value of $\vartheta_{,}+\varepsilon$; which latter indicates the most favorable angle for ricochet fire. By extending the operation we can readily ascertain between what values of $\varphi_{,}-\varepsilon$ the values $\vartheta_{,}+\varepsilon$ are less than 8°, and hence the range of angles of elevation respecting the horizon at which the shots will ricochet. By gradually increasing the assumed height of battery a limit is finally reached, at which the minimum value of $\vartheta_{,}+\varepsilon$ is about 8°. This marks the extreme height from which rebound is possible.

By constructing a series of curves (see Fig. 27, Plate I,) upon a system of co-ordinate axes of which the abscissas denotes the elevations referred to the horizon ($\varphi_{,}-\varepsilon$) and the ordinates the corresponding values of the angles of fall upon the water surface ($\vartheta_{,}+\varepsilon$), this discussion is rendered extremely simple. The curves thus constructed seem to be hyperbolic in character. The angle of elevation most favorable to ricochet in the particular problem under discussion lies at the point of intersection of the curves corresponding to different heights, and a curved line starting at zero on the axis of abscissas and inclining toward the depression side so that its cord makes an angle of about 13° with the axis of ordinates. From this diagram—after making certain allowances for the variations in initial velocity of the different shots—the following table of data, necessary for the solution of the proposed problem, has been derived. It should be remembered that it applies to the 15-inch gun fired with forty pounds of No. 5 powder, and a loaded shell weighing three hundred and forty-four pounds, giving an initial velocity of one thousand one hundred and sixty-six feet per second.

Data for determining ricochet trajectories at certain heights above water surface for gun, charges, &c, above specified.

Height of gun above water.	Angle of elevation referred to horizon. $\phi_{,}-\varepsilon$			Corresponding to value of $\phi_{,}-\varepsilon$, best for rebound.		
	Greatest for rebound.	Least for rebound.	Best for rebound.	Angle of fall upon water. $\theta_{,}+\varepsilon$	Striking velocity. $v_{,}$	Range, (initial velocity of 1166 feet.) $X_{,}$
Feet.	° ′	° ′	° ′	° ′	*Feet.*	*Yards.*
10	+5 30	—6 10 ?	—0 02	1 17	1060	300
36	+5 10	—6 10 ?	—0 18	2 41	1030	430
60	+4 30	—6 10 ?	—0 35	3 33	1000	550
104	+3 50	—6 00 ?	—0 49	4 50	960	730
150	+2 50	—5 50	—1 00	5 59	910	890
200	+1 30	—5 00	—1 25	6 54	870	1000
250	—1 20	—2 40	—1 55	7 48	840	1050

The computations upon these data are simple, and for convenience of reference will be stated in full. The formulæ used are the following—based, it will be remembered, upon the supposition that the resistance of the air may be allowed for in the numerical values of e and f and hence, that $\vartheta_{n}=\varphi_{n}$. *Eq.* (13) is deduced from *eq.* (1′). The quantity g is assumed at 32.18 feet. The quantities e and $f_{\prime\prime}$ have, of course, their values for shells of the same density as those used in the experimental firing.

(—) $e=1.05.$

(12). $f_{\prime\prime}^{2}=\sqrt{0.1153+13260\left[\cos(\vartheta_{\prime}+\varepsilon)-0.99\right]^{2}}-0.3395.$

(11). $f_{n}^{2}=f_{\prime\prime}^{2}+\dfrac{1-f_{\prime\prime}^{2}}{0.99-\cos(\vartheta_{\prime}+\varepsilon)}\left[\cos\vartheta_{n-1}-\cos(\vartheta_{\prime}+\varepsilon)\right]$

(5″). $V_{n}=f_{n}v_{n-1}\cos\vartheta_{n-1}\sqrt{\dfrac{e^{2}}{f_{n}^{2}}\tan^{2}\vartheta_{n-1}+1}.$

(6). $\operatorname{Cot}\varphi_{n}=-\dfrac{\cot\vartheta_{n-1}}{e}.$

$$(7).\qquad X_n=\frac{\sin 2\varphi_n V_n^2}{32.18}.$$

$$(1').\qquad y=x\tan\varphi_n-\frac{16.09\,x^2}{V_n^2\cos^2\varphi_n}.$$

$$(13).\qquad x=\frac{V_n^2\cos^2\varphi_n\tan\varphi_n}{32.18}\pm\sqrt{\frac{V_n^2\cos^2\varphi_n\tan\varphi_n}{32.18}-\frac{V_n^2\cos^2\varphi_n}{16.09}\,y}.$$

By means of *eqs.* (12), (11), (5″), (6), and (7), the lengths of the successive bounds, and the initial velocity and initial angle of elevation for each, are computed for each height of battery. The number of bounds in the general case is indicated by the number of values of f_n^2 greater than zero, and less than unity given by *eq.* (11.) It is, however, to be remarked that when the gun is near the level of the water, some of the last values of f_n^2 become so nearly unity that $X_n > X_{n-1}$, owing to the value of $\sin 2\,\vartheta_n$ increasing more rapidly than V_n^2 diminishes in *eq.* (7). Although such a result is theoretically possible, it seems hardly probable; and it has therefore been accepted as an indication that the final value of f_n^2 has been reached, *i. e.* that the shot sinks. *Eq.* (1′) may be applied to any branch, except the first where the resistance of the air must enter as a separate coefficient, by using the proper deduced values of V_n and φ_n. Hence substituting these values for each rebound in turn, and making $x=\frac{1}{2}X_n$ the corresponding values of y, will show the greatest heights attained above the water surface.

The only other quantity necessary to a comparison of these trajectories among themselves derives its importance from the fact that shot fired from elevated sites may rise in some of their rebounds so high above the water surface as to pass entirely over a hostile fleet. What height should be considered the maximum that is serviceable depends upon the height of the hulls of vessels of war. In reply to a letter requesting information on this point, Captain H. A. Wise, chief of Bureau of Ordnance, Navy Department, informs me that "the height of hulls of an unarmored fleet varies from twelve to twenty-five feet." Making y successively equal to twenty and twenty-five feet in *eq.* (13.) the corresponding length of each high branch dangerous to hulls of these heights has been computed. The whole branch is of course dangerous where these heights are not attained.

The following table exhibits the results of these computations as applied to the data contained in the last table:

Elements of best ricochet trajectories of 15-*inch gun, fired at certain heights above water surface, with charges above specified.*

Trajectory.	Gun 10 feet above water level.						Gun 36 feet above water level.					
	Initial velocity.	Initial elevation above horizon.	Greatest height above water.	Length within 20 feet of water.	Length within 25 feet of water.	Total length.	Initial velocity.	Initial elevation above horizon.	Greatest height above water.	Length within 20 feet of water.	Length within 25 feet of water.	Total length.
	Feet.	° ′	*Feet.*	*Yds.*	*Yds.*	*Yds.*	*Feet.*	° ′	*Feet.*	*Yds.*	*Yds.*	*Yds.*
From gun..	1166	—0. 20	10. 0	300	300	300	1166	—0. 40	36. 0	172	229	430
Rebound 1	968	1. 21	8. 1	457	457	457	887	2. 49	29. 5	346	487	799
2	884	1. 25	7. 4	400	400	400	765	2. 57	24. 1	367	623	623
3	807	1. 29	6. 8	350	350	350	661	3. 06	19. 9	489	489	489
4	738	1. 33	6. 2	305	305	305	573	3. 15	16. 4	386	386	386
5	675	1. 38	5. 7	269	269	269	499	3. 25	13. 7	306	306	306
6	617	1. 43	5. 3	236	236	236	435	3. 35	11. 5	245	245	245
7	564	1. 48	4. 9	207	207	207	381	3. 46	9. 7	197	197	197
8	517	1. 53	4. 5	182	182	182	335	3 57	8. 3	160	160	160
9	473	1. 59	4. 2	160	160	160	295	4. 09	7. 1	131	131	131
10	434	2. 05	3. 9	141	141	141	262	4. 21	6. 1	108	108	108
11	398	2. 11	3. 6	125	125	125	233	4. 34	5. 4	90	90	90
12	365	2. 18	3. 3	111	111	111	209	4. 48	4. 8	76	76	76
13	335	2. 25	3. 1	98	98	98	189	5. 02	4. 3	64	64	64
14	308	2. 32	2. 9	87	87	87	171	5. 17	3. 9	56	56	56
15	283	2. 40	2. 7	77	77	77	156	5. 33	3. 5	49	49	49
16	261	2. 48	2. 5	69	69	69	144	5. 50	3. 3	43	43	43
17	241	2. 56	2. 3	61	61	61	133	6. 07	3. 1	39	39	39
18	222	3. 05	2. 2	55	55	55	125	6. 25	3. 0	36	36	36
19	205	3. 14	2. 1	49	49	49	118	6. 44	3. 0	34	34	34
20	190	3. 24	2. 0	44	44	44	113	7. 04	3. 0	32	32	32
21	176	3. 34	1. 9	40	40	40	109	7. 25	3. 1	32	32	32
22	164	3. 45	1. 8	36	36	36	106	7. 47	3. 2	32	32	32
23	152	3. 56	1. 7	33	33	33	......	......	......	......	......	....
24	142	4. 08	1. 6	30	30	30	......	......	......	......	......	....
25	133	4. 20	1. 6	28	28	28	......	......	......	......	......	....
26	125	4. 33	1. 5	26	26	26	......	......	......	......	......	....
27	117	4. 47	1. 5	24	24	24	......	......	......	......	......	....
28	111	5. 01	1. 5	22	22	22	......	......	......	......	......	....
29	105	5. 16	1. 4	21	21	21	......	......	......	......	......	....
30	100	5. 32	1. 4	20	20	20	......	......	......	......	......	....
31	95	5. 48	1. 4	19	19	19	......	......	......	......	......	....
32	91	6. 05	1. 4	18	18	18	......	......	......	......	......	....
33	88	6. 23	1. 5	18	18	18	......	......	......	......	......	....
34	85	6. 42	1. 5	17	17	17	......	......	......	......	......	....
35	83	7. 02	1. 6	17	17	17	......	......	......	......	......	....

Table—Continued.

Trajectory.	Gun 60 feet above water level.						Gun 104 feet above water level.					
	Feet.	° ′	*Feet.*	*Yds.*	*Yds.*	*Yds.*	*Feet.*	° ′	*Feet.*	*Yds.*	*Yds.*	*Yds.*
From gun..	1166	—0. 50	60. 0	118	152	550	1166	—1. 10	104. 0	83	105	730
Rebound 1	808	3. 44	43. 0	236	310	879	666	5. 04	53. 7	168	217	807
2	657	3. 55	31. 3	243	336	610	475	5. 19	30. 1	182	254	431
3	538	4. 07	23. 2	270	430	430	349	5. 35	18. 0	245	245	245
4	444	4. 19	17. 4	307	307	307	266	5. 52	11. 4	148	148	148
5	370	4. 32	13. 3	223	223	223	209	6. 09	7. 8	96	96	96
6	311	4. 46	10. 4	166	166	166	170	6. 27	5. 6	67	67	67
7	264	5. 00	8. 2	125	125	125	143	6. 46	4. 4	49	49	49
8	226	5. 15	6. 7	97	97	97	124	7. 06	3. 7	39	39	39
9	196	5. 31	5. 5	76	76	76	112	7. 27	3. 3	33	33	33
10	172	5. 47	4. 7	62	62	62	105	7. 49	3. 2	31	31	31
11	153	6. 04	4. 1	51	51	51	102	8. 12	3. 3	30	30	30
12	138	6. 22	3. 7	44	44	44	......	......	......	......	......	....
13	127	6. 41	3. 4	38	38	38	......	......	......	......	......	....
14	118	7. 01	3. 2	35	35	35	......	......	......	......	......	....
15	111	7. 22	3. 2	33	33	33	......	......	......	......	......	....
16	107	7. 44	3. 2	32	32	32	......	......	......	......	......	....

Trajectory.	Gun 150 feet above water level.						Gun 200 feet above water level.					
	Feet.	° ′	*Feet.*	*Yds.*	*Yds.*	*Yds.*	*Feet.*	° ′	*Feet.*	*Yds.*	*Yds.*	*Yds.*
From gun..	1166	—1. 25	150. 0	66	80	890	1166	—1. 40	200. 0	57	72	1000
Rebound 1	493	6. 17	45. 3	138	181	548	325	7. 14	26. 0	142	219	274
2	305	6. 36	19. 1	220	220	220	197	7. 35	10. 5	105	105	105
3	211	6. 56	10. 1	110	110	110	154	7. 57	7. 0	67	67	67
4	163	7. 17	6. 6	69	69	69	144	8. 21	6. 8	62	62	62
5	137	7. 39	5. 2	52	52	52	......	......	......	......	......	....
6	125	8. 02	4. 8	45	45	45	......	......	......	......	......	....

Trajectory.	Gun 250 feet above water level.					
	Feet.	° ′	*Feet.*	*Yds.*	*Yds.*	*Yds.*
From gun	1166	—1. 55	250. 0	50	63	1050
Rebound 1	152	8. 11	7. 2	67	67	67

This table gives all the data necessary for a careful study of the loss in ricochet fire resulting from placing the battery at increased heights above the water surface. It is at once evident that at *no height* is the whole trajectory useful against a fleet. Fired from high sites the shot may pass over the vessels, and fired from either high or low sites, it soon loses the velocity necessary to be formidable to anything larger than a boat. What the limits of serviceable velocity and height are, is a matter of opinion. Assuming them to be three hundred feet and twenty-five feet respectively, which certainly verge upon the limits of utility, the following table, prepared from the one just given, shows the useful lengths of the longest ricochet trajectory for each height. The third column shows these quantities for four hundred feet, and twenty feet respectively. Other values can readily be computed.

Comparison of useful ricochet trajectories of 15-inch gun with charges above specified.

Height of gun.	Serviceable length of trajectory $Y_n < 25$ and $V_n > 300$.	Serviceable length of trajectory $Y_n < 20$ and $V_n > 400$.	Total length of trajectory.
Feet.	*Yds.*	*Yds.*	*Yds.*
10	3428	3007	4152
36	3122	2311	4457
60	1924	1174	3758
104	821	433	2706
150	481	204	1934
200	291	57	1508
250	63	50	1117

The data assumed for this problem are believed to be fair in every particular. Considering the solution exhibited by this table, and the facts that in windy weather, when the sea is rough, very little dependence is to be placed upon ricochet fire; that it is worthless against iron-clads; and that low sites exact enormous cost for plating, it would appear that where the site will permit a certain elevation, it is very possible to place a water battery too low.

The new formulæ are useful in discussing other points of interest in connection with the problem of defence, but it is not considered advisable to extend their application here.

CHAPTER III.---RIFLED GUNS.

Calibres employed in Virginia. The guns and carriages. More perfect sights essential to efficient firing at ordinary ranges. Detailed description of projectiles in use in both armies. Same of fuzes, including the English Armstrong 12-pounder fuzes. Ranges of 6-pounder James's rifle. Precision of fire with troops. Penetration of rifled projectiles of various calibres. General conclusions resulting from our experience with rifled guns in Virginia.

During the siege of Richmond no attempt was made at breaching or levelling the enemy's works by artillery fire. A few experiments in destroying the abatis in front of the works failed utterly, except at points where an enfilading fire could be employed. The rifled siege guns were therefore wholly used in batteries of position to keep down the confederate fire, annoy their working parties, interfere with their use of the Petersburg bridges, and repel or support assaults. On James river they played an important part in assailing the confederate fleet, whenever it ventured within range, and in protecting the digging of Dutch Gap canal.

The only classes of rifled guns in the siege train were the 100-pounder, 30-pounder, 20-pounder, and 10-pounder Parrott, (calibre 6.4, 4.2, 3.67, and 2.9, subsequently 3-inches;) the 4.5 and 3-inch Ordnance guns; and an experimental 24-pounder, (5.82 inches,) and 6-pounder (3.67 inches) rifled on the Sawyer plan. Of these the 24-pounder Sawyer gun burst on its tenth round, and guns of calibre less than the 30-pounder were scarcely used, being chiefly held in position to repel or aid assaults. We, therefore, really tested only the 100-pounder and 30-pounder Parrott, and the 4.5-inch Ordnance gun.

At Fort Fisher, and in Richmond and its batteries, we became possessed of many novel specimens of confederate ordnance, which were generally shipped to the arsenals, or transferred to ordnance officers, before any drawings of them could be made.

On the lines of Richmond their best rifled ordnance consisted of the 8-inch Brooke rifle, weight 22,000 pounds; the 7-inch Brooke rifle, weight 14,500 pounds; the old 32-pounder smooth-bore, (6.4 inches) banded and rifled, weight 8,000 pounds; the 4.62-inch siege rifle, weight 6,170 pounds; the 4.2-inch Brooke rifle, weight 4,700 pounds, and the 3-inch field rifle, weight about 1,100 pounds. They had, however, several other calibres. By Figs. 24 and 25, Plate I, the dimensions of their standard siege guns are shown.

GUNS, CARRIAGES, ETC.

For all our guns the rifling was shallow with nearly equal lands and grooves, thus permitting the use of round shot and shell if necessary. In the confederate service a greater diversity existed. Some guns were like ours; others were rifled on the Brooke's saw system, and occasionally the same as modified by Blakely; others were on Commander Scott's centrical system; others on the Whitworth system; others on the Armstrong shunt system. The confederate use of foreign ordnance caused this variety.

Beyond question, our 30-pounder Parrott gun was preferred to the 4.5-inch Ordnance pattern by battery commanders who had used both. This was partly because the latter was never fully tested before adoption, the trial gun having burst after about 800 rounds, thus throwing doubt upon its endurance; but chiefly because the projectiles supplied for it were, as will soon be shown, decidedly inferior to those of Parrott. The Ordnance model is much the handsomer, and more convenient to handle, and it has 750 pounds advantage in weight. Still, the fact as to how it was regarded is as stated. One minor defect in the Ordnance gun, which might easily be remedied, increased this prejudice; the gun is not bouched. A few rounds therefore cause a dangerous enlargement of the vent, as will be seen by examining Fig. 19, Plate I, which represents in plan and elevation a cast of the lower end of the vent of a 4.5-inch Ordnance gun that had been fired only 467 times. For these reasons the Ordnance guns were used much less before Richmond, and with more caution, than the Parrott guns.

Of the great endurance of the 30-pounder, and of the smaller calibres of Parrott guns, there is no dispute. One of the former was fired at Charleston, often at about 40° elevation, 4,606 times before bursting. None of those in my train were injured, except one, of which about a foot of the muzzle was blown off by the premature explosion of a shell in the bore. The new face of the piece near the bore was cut smooth with a cold chisel, and the accuracy of the gun seemed not to be impaired.

Against the endurance of the 100-pounder Parrott, much has been said and written. It is doubtless relatively weaker than the smaller calibres, but so far as our limited experience goes, not dangerously so. Our guns were always fired with special care to observe all conditions of safety. The bore was kept clean by bristle sponges and frequent washing, and by carefully excluding dust; the projectile was greased; the condition of the fuze plug was examined; the gun was run gently into battery so as not to start the shot forward in the bore, (a precaution not always observed in the navy,) and a careful record of the number of shots fired by each piece was kept.* We had no accidents with the gun, and it

---

* Learning, since this was written, that a series of experiments had recently been conducted at Mr. Parrott's foundry to test certain theories respecting the bursting of guns, I wrote requesting his present views upon this subject. The following letter was received in reply:

WEST POINT FOUNDRY,
October 12, 1867.

MY DEAR SIR: I have just received your favor of the 9th instant, and, in accordance with your request, I reply with pleasure to your inquiries as to any modification of my views concerning the danger to my rifle guns from premature explosion of shells within the bore of the piece, and also from the fact that the projectile is not "home" when the gun is fired.

As to the subject first mentioned, I would state that I cannot doubt the danger of bursting a rifle gun is much increased, and its subsequent endurance impaired, by the explosion of the shell within the bore; but that it is not necessarily the cause of fatal injury to the gun. This injury may in some cases be imperceptible, but in others it has been distinctly indicated by the sudden enlargement of the bore and the marks of the grooves upon pieces of the shell;

possessed great accuracy and penetrative power; indeed it was decidedly a favorite.

It was an object to divide the firing among the different guns as much as possible, pieces even being transferred from one battery to another occasionally for this purpose,

---

and many cases of the blowing off of parts of the chase, as with the 300-pounder at Charleston, are certainly traceable to the premature explosion of shells.

I have known, however, in the firing of one thousand rounds from a 100-pounder rifle gun, eleven shells to explode within the bore without bursting the gun; and yet in one instance (the only one, to be sure) the muzzle of the gun was carried away at the very first discharge by the explosion of the shell.

It is well known that with the same conditions, as to nature of powder and proportion of charge to projectile, the action of the powder becomes more severe as we increase in calibre. The gases formed by the combustion of the powder assume a higher tension, inasmuch as the mass to be moved increases faster in weight than the surface upon which the propelling power must act.

On the ignition of powder, the tension of the gas will depend on the weight of the projectile directly, and the surface on which the pressure is exerted inversely; and, from the great weight of the elongated or rifle projectile compared with the surface acted upon, the tension of the gas and consequent strain and liability to accidents to the projectile and the gun will be greater in rifle than in smooth-bore ordnance.

Thus, in the 100-pounder rifle the weight of the projectile is 3.1 pounds per square inch of the area of the bore, while in the 15-inch gun, with solid shot of 450 pounds, the same proportion is but about 2.7 pounds.

These facts do, I think, point to the more careful consideration of the powder for use in the larger rifle guns.

Somewhat connected with this point is the consideration of the effect of the repeated action upon nearly the same part of the bore by the expanding part of the projectile, unquestionably becoming very much more severe as it takes place more suddenly by the impulse of very quick powder.

I need hardly allude to the peculiar dangers with a sliding projectile arising from any obstruction in the bore, such as sand and the accumulation of dirt. In my judgment, the considerations which have led to the adoption of a peculiar powder in the large smooth-bore Rodman cannon, with such remarkable success, apply with even greater force to my large rifle guns. That the smaller guns, say 30-pounders and all lower calibres, have endured better than the heavier, is not peculiar to my guns, but follows from the very nature of the subject, and presents a difficulty which can be remedied by the precautions taken in other cases.

In respect to your other inquiry, it was found by direct trial that a space of even seventy-five inches between the charge of powder and the projectile did not cause the bursting of the gun. The effect was not apparently to injure the gun or the projectile, but merely to diminish the range.

Very truly yours,

R. P. PARROTT.

General H. L ABBOT.

in order to keep all in an equal condition of efficiency. For this reason no very large number of rounds were fired from any single piece. The following is the record of a few of the guns which did the most service:

100-p'dr	Parrott	No.	11,	fired	316 times.		All old guns fired an unknown number of times before coming into my possession.
"	"	"	13,	"	551	"	
"	"	"	15,	"	318	"	
"	"	"	20,	"	482	"	
30-pd'r	Parrott	"	100,	"	1,279	"	Apparently uninjured.
"	"	"	101,	"	1,578	"	
"	"	"	255,	"	1,585	"	
"	"	"	256,	"	1,631	"	
"	"	"	259,	"	1456	"	
4.5-inch	Ordnance	"	41,	"	467	"	Vents dangerously enlarged from not being bouched when made.
"	"	"	89,	"	578	"	
"	"	"	96,	"	509	"	
"	"	"	97,	"	529	"	

Only two serious imperfections in the rifled guns and carriages were developed by our firing.

The first is in the pointing apparatus, which is radically defective. For a sharpshooter's rifle, which is not expected to fire more than five or six hundred yards, we supply accurate globe sights and a really fine telescope. For a rifled gun, which is to fire three thousand or four thousand yards, we give sights far coarser than those of any old smooth-bore musket. The rim-base sight should be a delicate globe sight, covered by a cap for security when not in use. On the breech sight a slide should be arranged to carry, when needed, a small but good telescope provided with cross hairs. When not needed, this might advantageously be replaced by a sight consisting of a hollow double cone with a very small hole at the vertex;—the axis of the cone being placed parallel to the axis of the gun, and the inner surfaces being blackened to prevent reflections. The small globe sight seen through the small hole at the vertex of the cone gives every facility for accurately aiming at objects within the range of the unassisted vision. This peculiar conical eye-piece was the most noteworthy feature of a patent breech sight which the inventor brought to me for trial. His sight was too complex for use by soldiers, but on trial by Major Hatfield it decidedly improved the practice of the 100-pounder Parrott. A simple arrangement, carrying out the idea of the combined globe and breech conical sight, would be of great utility. An ordinary breech sight should also be in the gunner's haversack for use in quick firing at short ranges. It may be objected that these sights would be expensive and too delicate for the field. The reply is that they should be as strongly made as possible, and that care corresponding to their value must be used in handling them. They are *absolutely essential to accurate firing at long or even ordinary rifle ranges.* I have repeatedly used my telescope to detect an object quite invisible to the naked eye, and by aiming at its position, as indicated by surrounding larger objects, have made practice impossible for a gunner not supplied with a telescope. This method of sighting is of course very laborious and uncertain, but it proves that with our large rifled guns the unavoidable uncertainties of the trajectory are exceeded by the errors inherent to our present system of pointing. We place telescopes on surveying instruments and on the rifles of sharpshooters; why not on the rifled siege gun, the ammunition of which is too expensive, too heavy, and of too much importance in the campaign, to be wasted? Rifled artillery of the larger calibres can never accomplish what it ought until this matter receives attention.

The second defect is in the elevating arrangement. As the Parrott guns are now made without preponderance, the elevating screw is designed to depress as well as to elevate the piece. It is not sufficiently strong at the shoulder, which soon gives way. The 30-pounder guns were often fired in the batteries without the screw; but this is objectionable, partly because it is exceedingly difficult to give a delicate motion by hand, and partly because without the screw the guns when discharged sometimes rotate on the trunnions with considerable force, a result due probably to their axis not intersecting *exactly* the axis of the piece.

All banded guns are exposed to one danger so remote as to be hardly worth considering, viz., that of becoming suddenly and effectually spiked by the slipping back of the band from the impact of a hostile shot. This actually happened to a 7-inch Brooke gun in the confederate James river batteries. The gun was struck by one of our shots, and thus spiked when loaded, and was transported to the Tredegar works for repairs in that condition. It was pointed out to me by one of the watchmen of the works after the occupation of the city. It had been necessary to remove and reset the band.

The siege-gun platforms issued by the Ordnance department are designed for breaching batteries where the direction of the gun is but slightly changed. For batteries of position they are entirely too small. We made very little use of them, preferring to improvise a larger platform of lumber obtained from neighboring buildings.

RIFLED PROJECTILES.

Rifled cannon have been introduced by the present generation. Indeed, their first really successful application to actual warfare occurred in Napoleon's Italian campaign of 1859. Heretofore, the difficulty seemed to be to discover any projectile which could be made to take the rifled motion without injury to the gun. To-day the difficulty is to select the best among many of such projectiles. During the late war both belligerents devoted great attention to this matter. Many inventions brought to the test of actual warfare were laid aside, but so gradually that it is difficult even now to state the ideas upon the subject entertained at any given date. Careful collections of all confederate projectiles fired at our batteries in 1864–'65 were made and forwarded to my headquarters, where they were examined and accurately drawn. Rough field photographs of the collection distributed among the troops excited the interest of both officers and men; and I am satisfied that at the end of the siege my collection, enriched in the larger calibres by the spoils of the Fort Fisher and Richmond magazines, was essentially a complete exhibit of the confederate views and experiments in this branch of artillery. In connection with our own projectiles in actual use at that time, they give a definite idea of the systems of rifled artillery employed in America at the close of the war. Plates representing the collection in detail have been prepared to accompany this paper. The original projectiles are deposited in the artillery museum at West Point.

It is noticeable that, while neither belligerent copied the other in details, there was a strong general resemblance in the artillery most in use. This was doubtless due in part to the fact that Messrs. Parrott and Reed, whose inventions were much used in the two armies, had experimentally studied the subject together before the war. Breech-loading cannon were discarded, as well as the now nearly universal European system of studded or flanged projectiles. A few imported guns of the Whitworth, the Blakely, and the Armstrong patterns were employed by the confederates; but the American rifled artillery of both armies consisted almost exclusively of muzzle-loading guns with expanding projectiles.

In our service, experience had caused the rejection of the James and of other varieties of rifled projectiles before the end of the war. For the siege of Richmond we used (except experimentally) only those of Parrott, Schenkl, and Hotchkiss. Of these, the first were used with the Parrott guns, and the last two with the Ordnance guns. Case shot and shells were made on all these systems, and solid shot on all but that of Schenkl. The guns were also supplied with canister not designed to take the rifled motion.

The *Parrott projectile* (Figs. 1, 2, 3, 4, 10, 11, 15, 20, Plate II) takes the grooves by the expansion of a ring of brass attached to its rear end. Earlier in the war the 10-pounder and 20-pounder projectiles had wrought-iron rings or cups for this purpose; but experience caused them to be rejected for the brass rings, which in the smaller calibres project very slightly beyond the base of the shell, forming a small cup-like cavity or recess to facilitate the expansion. In the larger calibres this recess is found to be unnecessary, but the ring should be slightly started with a cold chisel in the battery at three or four points of the circumference—unless already started before issue, as was Mr. Parrott's latest practice with all calibres.

The Parrott projectile was deservedly the favorite in the service. About ninety-five per cent. of those fired in the siege of Richmond took the rifled motion well. It had, however, two defects. First, the brass ring often separated from the iron at the discharge, and, flying in fragments in front of the battery, exposed our own advanced troops to danger. Mr. Parrott attributes this to variation in quality of the powder; which renders it impossible to guard against too great rigidity on one hand, and deficient strength on the other. And second, the inner surface of the shell as issued was rough; which occasionally, but very rarely, caused premature explosions from attrition, thus endangering both the gun and the troops in front. (Another theory for these explosions was that they were caused by defects in the casting at the base of the shell, which admitted the flame to the bursting charge.) The larger calibres were more liable to this danger, and Mr. Parrott, in a letter, suggested to me the use of a preventive coating composed of thirty-two parts, by weight, of common bar soap, twelve of tallow, and ten of resin. These ingredients are to be melted together, and, when very fluid, poured into the shell in sufficient quantities to thoroughly coat the whole interior surface. This coating is rendered uniform by rolling the shell about, and the surplus is then poured off. Mr. Parrott stated that, after many trials, he had found this preparation effectual in preventing premature explosions; but it was never tested before Richmond, because these explosions were so rare in our calibres that it was deemed an unnecessary labor.

The *Schenkl projectile*, (Fig. 6, 12, 16, 17, 18, Plate II,) takes the grooves by the forcing of a papier-maché sabot forward upon its slightly conical base at the instant of discharge. When the sabot is *well made* and in *good order*, this is excellent ammunition. It has a smoother and more silent flight than the Parrott, it gives excellent practice, and the light sabot does not endanger troops in front. A great practical difficulty, however, is found with the sabot, which is liable to swell from moisture, so that the piece cannot be easily loaded. To remedy this defect, Mr. Schenkl, early in the war, covered the papier-maché with thin zinc; but the expedient was a failure and had to be abandoned. Some of this old ammunition, designed for a 30-pounder Parrott, was issued for use by the siege train in 1864, and again proved itself worthless. Even cutting off the zinc did not prevent the shell from tumbling when fired from the Parrott gun, possibly in consequence of the gaining twist of the rifling. Our new ammunition of this patent issued for the 4.5-inch Ordnance gun was very inferior to that made prior to the death of the inventor. The sabot, instead of having a fibrous structure, was hard and brittle like a lump of clay. It was tried in every condition—dry, damp, wet, a little shortened at front end—but in vain. The sabots of 21,000 rounds were condemned and in part replaced, but without entirely obviating the difficulty. About twenty per cent. of the improved issue tumbled; which in firing over our pickets at long ranges endangered them, both from the shortening of the range, and from the occasional explosions of the percussion shells, caused by the rapid end-over-end rotation. In fine, we modified the high opinion of this ammunition formed from experiments with the earlier samples in the defences of Washington, where certainly it appeared superior to any other. When the sabot is *good* the shells seem to take a more uniform rotation than those of Parrott, which, judging by the sound, differ among themselves, some rotating sufficiently to preserve the longitudinal axis steady, others allowing it a slight wabbling motion, and others one so excessive as to cause absolute tumbling. These effects are doubtless due to variations in the powder, which, by its greater or less quickness, causes the grooves to take hold of the brass ring at different distances from the seat of the charge. The elastic toughness of the good Schenkl sabot communicates a rotary motion to the projectile at once, and thus obviates this cause of inaccuracy. It should be added, however, that it also gives a greater strain to the gun.

The *Hotchkiss projectile* (Figs. 14, 19, Plate II,) takes the grooves by the forcing of the rear iron cap forward at the instant of discharge, thereby compressing the lead band surrounding the middle part, where the iron portions overlap, and forcing it to expand into the grooves. This projectile, which closely resembles Lynall Thomas's English patent, was used considerably by the light artillery, but from the strain exerted by it upon the gun I did not like to employ it in the larger calibres. It was accordingly not tested for the siege train.

About the close of the war Mr. Hotchkiss patented a new projectile, which takes the rifled motion from the flattening of a cupped brass ring cast upon its base, and held firmly in position by an extension of the iron sides over and through it in three or four places; thus allowing a certain windage to reduce the strain on the gun. This projectile is designed for the larger calibres, and was never tested in the field.

In addition to these standard kinds of ammunition, experiments were made with other patents as follows:

The *Dyer solid shot and shell* (Figs. 8, 9, Plate II) take the grooves by the expansion of a leaden cupped sabot cast upon the base and rear half of the projectile, which is grooved, and to give greater hold bound round with a network of fine wire. Experiments have been continued with this projectile since the war, resulting, I am told, in decided improvements; but the pattern issued for use with the 4.5-inch Ordnance gun of the siege train was made as above described. Only about eighty per cent. took the grooves, and the sabot was generally torn off and thrown in fragments in front of the battery to the peril of advanced troops.

The *Absterdam projectile* (Fig. 7, Plate II) takes the grooves by the expansion of a leaden cupped sabot cast upon its base. To prevent wabbling and to give a smoother bearing upon the lands of the gun, the projectile is surrounded by two narrow rings of lead (indicated upon the drawing) held in grooves cut near its extremities. This projectile, for the 4.5-inch gun, proved an utter failure, as the following records show. The charge of the gun was 3.25 pounds.

Test of the Absterdam projectile.

Battery.	Commanding officer.	Elevation.	Number fired.	Took grooves.	Tumbled.	Uncertain.	Burst well.	Burst at muzzle.	Did not burst.	Uncertain.	Remarks.
		° ′									
Fort Brady.	Capt. Pierce.	2—00	1	1	......	....	1	....	....	....	Percussion.
		2—08	4	1	3	....	......	....	3	1	Do.
		3—10	1	1	......	....	1	....	....	....	Do.
		3—15	177	45	113	19	70	19	48	40	Do.
		4—45	41	8	28	5	22	5	10	4	Do.
Battery 17..	Lt. Rogers...	5—08	4	1	......	3	......	3	....	1	Do.
		5—30	8	7	......	1	5	....	1	2	Do.
Fort Morton	Lt. Patterson	4—30	2	1	......	1	1	1	....	....	Time fuze.
		4—45	1	....	......	1	......	....	....	1	Do.
	Total....	....	239	65	144	30	100	28	62	49	

The siege train was supplied with two *Sawyer guns* for the trial of that inventor's rifled ammunition. One was of calibre 5.8 inches, and the other of calibre 3.67 inches, and the ammunition differed entirely for the two. For the former (Fig. 5, Plate II,) the shell belonged to the flanged system. It was thinly coated with lead to lessen the wear upon the lands and grooves of the gun. The practice from this gun was good until the eleventh round, when it burst into four principal fragments. The largest, including the trunnions and all in front of them, remained in its place on the carriage ; the next piece, forming the bottom of the bore near the breech, fell between the cheeks; the left half of the top, which split as usual through the vent, lodged on the top of a return of the parapet a few feet from the gun; the right half was thrown many yards entirely outside of the battery. The vent was evidently defective, showing a much enlarged double cavity at its lower end, probably caused in bouching. The gun had been fired an unreported number of times at Fort Monroe, and its endurance of the strain of this projectile being doubted, it was fired by quick match,—a precaution which saved the cannoniers from injury. The projectile for the 3.67-inch Sawyer gun is shown on Fig. 13, Plate II. It takes the grooves by the expansion of a leaden sabot attached to the rear half of the projectile, and bound round with iron foil. According to our limited experience about seventy-seven per cent. seemed to take the grooves. The leaden sabot often stripped.

Two patterns of *incendiary shells* were tested by the siege artillery brigade ; one, that of Mr. Fleming, (Fig. 11, Plate II,) the other, that of Mr. Berney, (Fig. 2, Plate II.) They both consisted of the ordinary Parrott shell with its cavity sub-divided by a thin diaphragm, perpendicular to its axis, into two compartments. The front compartment contained the bursting charge and the rear compartment the incendiary composition, introduced through a hole in the base of the shell subsequently closed by a screw plug and copper washer. Both parties claimed the invention, and both used an incendiary fluid of secret composition. By both the rear compartment of the shell was charged with raw cotton, or other similar substance, and was then filled with the fluid. The two inventions were entirely identical; except, perhaps, in the composition of the incendiary fluids, which being kept secret by both parties, may or may not have differed.

Mr. Berney's projectiles were received in October, 1864, and in the following month were fired at certain wooden houses within the enemy's lines by Captain Pierce, 1st Connecticut artillery, commanding Fort Brady, on the north side of James river. The projectile was the long one hundred pounder shell, (calibre 6.4-inches,) containing about six pints of the liquid. Six houses were consumed, at a range of one thousand five hundred yards, by the expenditure of twenty-nine shells. Forty-two of these shells were fired—without any premature explosion or other accident. Whenever the shell burst within a house a conflagration ensued. The fluid poured upon the ground burned about five minutes with a clear flame of low temperature. It apparently contained turpentine, and petroleum. The inventor claimed that it would ignite at a temperature of 120° Fah., and would consume in a shell without explosion if ignited by hand; but that the ignition of the bursting charge made it explode violently by the great heat engendered.

Mr. Fleming's projectiles were received in November, 1864*, but remained untouched until March, 1865, when they were tried before a board of officers convened by order of General Grant. The samples furnished were for the 30-pounder Parrott gun, calibre 4.2 inches. The bursting charge was 4.5 ounces of musket powder. About a pint of the liquid was used in each shell. Fifteen shells were fired with a range of about three hundred yards, at a palisading, an abatis, and a pile of logs—the materials of which had been cut about four months. A projectile burst well in each of these targets, but failed to ignite them. One shell, however, set some dry leaves and brush on fire. The contents of a shell poured upon the ground and ignited burned for five minutes, but with a flame possessing very little heat. A tin can containing five quarts of the liquid was then covered with brush and ignited. It exploded at once and consumed the brush. A second can exploding like the first, under the abatis, burned out without consuming the larger branches. The invention was entirely useless for burning abatis, stockading, or palisading, unless surrounded by dry leaves or other very combustible material. The projectile would doubtless be better than the common

* Notification of their having been ordered was sent to me by the Chief of Ordnance on September 11, 1864.

shell for firing houses or shipping, although it is doubtful whether the 30-pounder shell would contain enough of the liquid to accomplish these ends. No premature explosion or other accident attended these experiments.

As already explained, advantage was taken of the expenditure of rifled ammunition before Richmond to institute, by means of careful records, a comparison between the different samples issued. Kept under the fire of the enemy and with graver matters demanding the chief attention, the records doubtless contain errors; but the reports from the different batteries so well accord among themselves that, after a close scrutiny of them, I am satisfied none of these errors are material; in fact, among the multitude of shots fired they are probably cancelled.

Service test of United States rifle ammunition.

Kind of gun.	Projectile.	Number fired.	Uncertain.	Number tested.	Took grooves.	Tumbled.	Percentage serviceable.	Remarks.
100-pounder Parrott	Parrott..........	1375	87	1288	1244	44	0.96	
30-pounder Parrott.	....do.	10901	1860	9041	8705	336	0.96	
Do..........	Schenkl, (banded)	178	56	122	70	52	0.57	
20-pounder Parrott.	Parrott..........	849	183	666	574	92	0.86	
Do..........	Schenkl.........	49	0	49	46	3	0.93	
3-inch Parrott.....	Parrott..........	40	6	34	28	6	0.82	
4.5-inch Ordnance.	Schenkl.........	3213	520	2693	2212	481	0.82	
Do..........	Dyer..........	1364	142	1222	981	241	0.80	
Do..........	Absterdam, (lead)	239	30	209	65	144	0.31	
3-inch Ordnance...	Hotchkiss.......	296	177	119	114	5	0.95	
5.82-inch Sawyer..	Sawyer, (flanged)	10	0	10	10	0	1.00	Gun burst.
3.67-inch Sawyer..	Sawyer, (lead)...	125	20	105	81	24	0.77	

This closes the subject of our own rifled projectiles. As already stated, our collection of confederate ammunition was very extensive, and chiefly obtained by gathering up what was fired into our batteries; of course we only learned what we could see respecting the different varieties. After the end of the campaign, I was unfortunate in not meeting any confederate artillery officer competent to give information either as to the names, history, or merits of some of the different patterns. All that can be done, therefore, is to present and classify the collection, and draw such inferences as seem to be legitimate respecting the merits of the different devices. The drawings are, in all cases, counterparts of the originals, no attempt being made to restore lost parts.

It is to be remarked that in calibre, in variety of pattern, and in systems of rifling, the confederate artillery was greatly more complex than our own. This is evident from the following general classification of the collection, which comprises ten distinct systems for giving rifled motion. Many of the differences between specimens of the same system are slight, but are nevertheless important and suggestive for future experimental research.

Classification of confederate ammunition.

Calibre.	Number of samples collected in the different systems for giving rifled motion.										Total number of samples.
	1st	*2d*	*3d*	*4th*	*5th*	*6th*	*7th*	*8th*	*9th*	*10th*	
64-pounder, 8 inch....	4	1	2	....	1	....	2	3	....	....	13
42-pounder, 7 inch....	1	3	7	1	2	1	....	....	....	....	15
32-pounder, 6.4 inch..	3	9	1	2	2	....	....	....	....	....	17
24-pounder, 5.82 inch.	....	1	....	....	....	....	....	....	....	....	1
18-pounder, 5.2 inch..	....	....	....	....	....	....	....	....	1	....	1
12-pounder, 4.62 inch	1	3	....	....	1	....	....	....	....	....	5
9-pounder, 4.2 inch..	6	....	2	....	....	....	....	....	....	1	9
6-pounder, 3.67 inch.	3	....	....	1	....	....	....	....	....	....	4
4-pounder, 3.4 inch..	....	....	....	2	....	....	....	....	1	....	3
3-pounder, 3.0 inch..	7	....	....	2	....	3	3	....	1	....	16
2.3 inch..	....	....	....	....	....	2	....	....	....	....	2
2.2 inch..	....	1	....	....	....	....	....	....	....	....	1
1.4 inch..	1	....	....	....	....	....	....	....	....	....	1
Total..............	26	18	12	8	6	6	5	3	3	1	88

One small but characteristic difference between these projectiles and our own at once strikes the eye. Usually the confederates only turned two raised rings on their shells, leaving the rest of the surface rough, while we smoothed the entire shell by forcing it through a die. Whether this was merely to save labor, or whether they considered that it reduced the wear upon the lands of the gun, is not known to me.

On the drawings the different projectiles are classified by calibres, but for description, as already intimated, they should be grouped according to the device employed for causing them to take the rifled motion. The ten systems will now be described in turn.

The *first and most common system* is Reed's, which closely resembles that of Parrott. The rotary motion is given by an expanding ring of soft metal attached to the base of the projectile—wrought iron, copper, and even lead being employed for this purpose, but no brass. The larger specimens are comparatively rare; eighteen out of twenty-six samples being for field and siege guns, for which it seems to be the preferred system. The plans for attaching the ring are numerous and well worthy of study. The drawings show ten different devices, some very ingenious, and one (Fig. 56, Plate VI) decidedly more successful than that of Parrott. The projectile was often thrown into our batteries, and some captured samples were fired from a 30-pounder Parrott at the enemy. So far as we could judge, it was free from the three great faults to which this system is liable, viz., failing to take the grooves, throwing off the ring, and chipping off dangerous fragments from the base of the shell. The troublesome use of the cold chisel for starting the ring is also avoided. In my judgment, this pattern ought to be thoroughly tested with a view to use in our own service. Some captured samples of Fig. 53, Plate VI, were also fired from our batteries, with fair results. It is interesting to see from Figs. 34 and 35, Plate V, that the confederates, like ourselves, found difficulty in making use of some patterns of copper rings and of wrought-iron cups; for, to render the ammunition serviceable, they were evidently obliged to discard the ring and employ another device. These samples were captured in the magazines of Fort Harrison, north of James river, and were designed for an old 32-pound sea-coast gun, banded and rifled.

The use of copper was general in the confederate service, not only for rings and plates on rifled shells, but also for fuze plugs, for which we generally employ brass. A knowledge of this fact was often useful in deciding the origin of doubtful samples of ammunition. Thus the device shown by Fig. 1, Plate III, is identical on the exterior with Parrott's, but having the copper ring, the shell was clearly from a confederate workshop.

To discover what kind of rifled field ordnance was most used by the confederates, what pattern of ring was found most serviceable by them, and how their ammunition endured the test of firing, Major General Warren, commanding 5th Army Corps, caused, as already stated, a collection of unexploded confederate shells to be made upon his portion of the Petersburg lines. This he turned over to me. The following is the record for the 10-pounder, the only rifled gun extensively used at that time and place by them:

Fig. 81, *Plate* VI.—Number of samples, 24—took groves well; lost ring, 3; iron chipped from base, 4.

Fig. 80, *Plate* VI.—Number of samples, 19—groove marks not well defined; lost ring, 1; iron chipped from base, 1.

Fig. 79, *Plate* VI.—Number of samples, 8—groove marks not well defined; lost ring, none; iron chipped from base, 4.

Fig. 82, *Plate* VI.—Number of samples 5—took grooves well; lost ring, none; iron chipped from base, 3.

Fig. 84, *Plate* VI.—Number of samples, 1; lost band, 1.

There was also nine samples of our Hotchkiss, and one of our Schenkl 10-pounder ammunition.

The *second device* is a curved copper plate secured by a screw and held firm by three dowels made usually of three copper projections from the plate extending into holes in the iron base of the shell; but sometimes of three iron projections from the base of the shell extending through holes in the plate. The explosion of the powder flattens the plate and thus gives the rifled motion by increasing its calibre. This system is liable to the objections that the plate almost invariably separates from the shell, rendering the projectile unfit to be used over troops; and also that it is very costly, from six to eight pounds of copper being used with a single shell of the larger calibres. It was, nevertheless, very common in the confederate service, chiefly for the larger calibres. The drawings show eighteen different patterns of shells to which it was applied; of which nine are for the 100-pounder (6.4 inches,) and four for larger guns. It is, however, applied to the diminutive shell shown in Fig. 87, Plate VI, which weighs only three pounds. From the fact that this system was used at the siege of Yorktown, and that it was not largely used with the more recent 7-inch gun, I infer that it was superseded by the pattern to be next described. Attention should be invited to the extraordinary shells represented by Figs. 38, 39, and 40, Plate V. They were found in the naval laboratory in Richmond, and it is not absolutely certain that they were designed for rifled guns, although I have little doubt upon that point. They were clearly intended for iron-clad warfare. The peculiar method of attaching the copper plate by friction is that designed by Stafford, strengthened, it will be noticed, by a square iron projection in the base of the shell according to the Blakely system, and in one instance by three small dowels in addition.

In the *third system* the rifled motion is given by means of a cupped copper plate termed "ratchet sabot," secured to the shell by a central screw and held firm by radial grooves (generally seven in number) in its iron base. This system was sometimes known under the name of "Tennessee Sabot." Two samples, Fig. 6, Plate III, and Fig. 18, Plate IV, bear Brooke's name upon the copper plate. From this circumstance, and from the fact that the system is largely used in his well-known 7-inch guns, I infer that he is its designer, or, at least, that it is favored by him. It seems to be exclusively confined to the heavier guns; out of twelve samples nine being for the 7-inch or larger calibres and none for smaller guns than the 30-pounder (4.2 inch.) The plates adhered to the shells much better in this system than in the one last described, but still they were frequently missing. In the matter of cost it is even more extravagant. Attention should be directed to Fig. 23, Plate IV, which represents a shell evidently designed from its great length for punching iron-clads. The device of placing the fuze-hole in the side is novel.

In the *fourth system* the rifled motion is given by the expansion of a lead sabot, which sometimes, but rarely, remains on the projectile. The method seems to be both costly and clumsy. The devices for attaching the lead are curious studies. Fig. 65, Plate VI, is clearly upon the system of Bashley Britten. Indeed, this shell was captured with one of Blakely's guns in Fort Clifton. The lead in some instances (as on Fig. 67, Plate VI) must have been attached with zinc solder by the Britten invention. In Fig. 77, Plate VI, a covering of tinned iron was firmly compressed upon the irregularities of the iron, and the lead cast upon that. The lead rarely adhered to the shell, but the tin generally remained.

The *fifth system* I believe to be original with the confederates. It consists in making the projectile (always solid) of wrought-iron bars coiled and welded. It is then accurately turned and a narrow groove cut around the base near the circumference so as to leave a thin ring, which is expanded by the action of the powder. It is rare and costly;—and, moreover, according to recent experiments, both steel and chilled iron prove better for punching iron-clads than wrought iron. These projectiles were designed exclusively for this use, although they were occasionally fired at the land batteries by the confederate fleet, and one (Fig. 51, Plate VI) from Fort Clifton, Appomattox river. They were made in the Tredegar works at Richmond, where samples in all stages of construction were found when the place was captured; also samples upset but not split, which had evidently been fired at iron targets. Fig. 28, Plate IV, is drawn from a specimen captured in Fort Fisher, North Carolina, and sharpened to a point for punching iron-clads.

The *sixth system* is that of Whitworth, viz., a projectile with hexagonal sides accurately fitted to a corresponding interior surface of the gun, slightly twisted to give the rifled motion. Both breech-loading and muzzle-loading guns are made on this system. The cartridge for the former is made of tinned iron, shaped to fit the bore and containing between the powder and projectile a lubricating wad of wax and tallow. These projectiles were largely used by the confederates on the lines of Petersburg, where they inspired dread among our men from their long range and horrid sound. Being solid shot they did us very little actual damage, but their great penetration in parapets and bomb-proofs, and the terrible wounds which they inflicted upon men and horses, invested them with a kind of mysterious terror; and at no time was the artillery so urgently called upon to silence the enemy as when the "Whitworth battery" was in action. The confederates were sensible that the want of shells was a disadvantage, and made a rough attempt (shown on Fig. 74, Plate VI) to supply the deficiency. The interior capacity, however, was so small that the number of fragments was limited to two or three. At Fort Fisher, North Carolina, several beautiful shells of small calibre, (Figs. 73 and 86, Plate VI,) evidently of foreign manufacture, were found; as well as the large shell, shown on Fig. 26, Plate IV, for which there was no corresponding gun;—but none of these patterns were ever used on the Petersburg lines. Two distinct kinds of solid shot, however, were there used; one elaborately finished and doubtless of foreign manufacture, the other rough and suggestive of confederate work shops.

It may appear strange that a gun so highly prized by the enemy was not found in our service. This was no doubt partly due to political considerations, but not wholly so.

A battery of six 12-pounder Whitworth guns was presented to the government by citizens abroad at the beginning of the war. At the siege of Yorktown, four of them, (and during the succeeding campaign of the Peninsula two of them,) accompanied the army, served by a detachment of the 1st Connecticut artillery. The slight difference between their practice and that of our own guns, their lack of shells, case shot, and canister, and the difficulty of supplying special ammunition to an army on the march, caused them to be returned to the Washington arsenal at the end of that campaign; and they were never afterwards put into the field. The other two guns of the battery were kept in position at Fort Ward, defences of Washington south of the Potomac, during the entire war; where a few experiments were made with them, as will be hereafter detailed.

The *seventh system* is that of Armstrong. His early breech-loading system is represented by the shell shown on Fig. 72, Plate VI, which was fired into one of my Dutch Gap batteries in the summer of 1864. Had it not been for this circumstance, I should have supposed that none of the ordnance of this gun-maker was ever used by either army in Virginia. Just at the close of the war, however, the confederates were supplied with an elegant and complete battery of muzzle-loading shunt 12-pounder Armstrong guns, but it was captured on the retreat from Richmond before it had fired a shot. The ammunition is shown by Figs. 70 and 71, Plate VI. The former is the celebrated segment shell, which serves for canister, case shot, and solid shot, according to the time at which its fuze is set. The latter is the shell for blowing up earth-works, firing magazines, &c. To be obliged to carry only two kinds of ammunition in the limber chests is certainly a great gain; but other peculiarities in this battery, such as the method of slightly traversing the trunnion beds, the singular arrangements of the limbers, the fuzes, &c., were not generally considered improvements upon our own system.

At Fort Fisher, North Carolina, the projectiles shown on Figs. 9 and 13, Plate III, were captured with the 150-pounder (8-inch) muzzle-loading shunt Armstrong gun there mounted. This gun was the most elegantly finished piece of artillery I ever saw. With its mate, captured in Fort Caswell, it no doubt constituted a valued present to the confederacy, whose waning fortunes transferred it to our hands. Rusted and its beauty gone, it is now preserved as a trophy at West Point. Of its two projectiles Fig. 13 represents the ordinary shunt shell, whose studs readily slide down the deep half of the grooves, to be turned into the shallow half, and then to be firmly nipped under the combined influence of the exploding charge and the twist of the rifling. The projectile shown on Fig. 9 could only have been intended for use upon iron-clads at very short ranges, where no rifled motion is necessary.

The *eighth system* consists of the three deep grooves in the gun and corresponding flanges on the projectile devised by Scott and adopted by Blakely for his large calibres. These projectiles (Figs. 11 and 12, Plate III) were captured at Fort Fisher, North Carolina, where the gun to which they belonged was the most efficient of the confederate armament, and was taken marked by our shot and stained with the blood of its cannoniers. Fig. 10, Plate III, was evidently designed for close quarters, where the rifled motion is unnecessary. There were several Blakely guns of smaller calibre found in and about Richmond, and one in Fort Clifton, on the Appomattox river; but they were designed for his lead-coated projectiles, and were very little used.

The *ninth system* is identical with that of Hotchkiss already described; but certain of the projectiles, as those shown on Figs. 47 and 68, Plate VI, were manufactured for or by the confederates. This is shown by the absence of the usual patent marks upon these shells, and by the facts that the former belongs to a calibre not used in our service, and that the latter has a stout wire wrapped around the projectile and imbedded in the lead. The system was considerably used, as one accustomed to its peculiar sound could easily detect by the ear when much exposed to confederate fire. Where or how they obtained their supply is open to conjecture.

The *tenth and last system* is identical with that of Schenkl already mentioned. This was very little used by the confederates. In fact, I have some doubt about the propriety of including it in the list, but do so because the shell shown by Fig. 61, Plate VI, was fired by them at one of my batteries. Its exterior form is identical with that of the old Schenkl 30-pounder shell, (Fig. 12, Plate II,) a few of which we had fired into their lines before this was received. It may be one of those shells, but it contained the characteristic copper-fuze plug of their service, and must have been supplied with some kind of confederate sabot.

FUZES.

The fuzes used with the guns of the siege train consisted of the Parrott and Dyer time; the Parrott, Schenkl, and Absterdam percussion; the Tice concussion; the Schenkl and Sawyer combination; and (for smooth-bore field guns and howitzers exclusively) the Bormann time fuze.

The Parrott and Dyer time fuzes consist, of a hollow fuze plug of soft metal screwed into the fuze hole of the shell; and of a slightly conical paper case made to exactly fit the hole of the plug, and filled with fuze composition prepared to burn for a greater or less time, according to the range desired. The time of its burning in seconds is marked on each paper case, and may be modified, if necessary, by shortening the latter before its insertion into the plug. These two fuzes are not essentially unlike, but the difference in the system adopted for giving rifled motion to the projectiles causes a slight difference in the percentage which fail to ignite.

The Parrott, Schenkl, and Absterdam percussion fuzes consist of a hollow fuze plug screwing into the fuze hole of the shell and containing a movable plunger. The interior hollow is partly closed at the bottom by a diaphragm pierced by a hole to admit the flame to the charge, and entirely closed at the top by a solid diaphragm, which may be unscrewed at pleasure. The plunger is hollow, charged, except in the Absterdam fuze, with powder, and provided at its front end with a nipple holding a common musket cap. The only essential difference between the three fuzes is in the manner of holding the plunger in its place to avoid accidental explosions. In the Parrott fuze this is accomplished by placing over it a hollow cylinder of soft metal, supported by two small projections resting on a ring near the front end of the plunger. These projections are broken by the inertia of the plunger at the instant of impact; and the latter passing through the cylinder, the cap is exploded against the front diaphragm. It was found that this soft-metal cylinder must be removed to insure ignition, unless it was designed to direct the shell against an object offering considerable resistance. If the plunger is at the bottom of the fuze, no premature explosions result from this removal. In the Schenkl shell the plunger is held in place by a small screw, which, passing through the side of the plug, holds it steady against any slight shock, but which is not strong enough to resist the inertia of impact. For additional security the top diaphragm is arranged with one counter sunk and one solid end. For transportation the former is screwed in next the plunger, but just before loading it is reversed. The screw sometimes prevents explosions, unless the target offers considerable resistance. The Absterdam plunger is held firm by a small hollow cone of lead joined at its apex to the base of the lead plunger. The base of the cone extends through the hole of the lower diaphragm opening out to a larger diameter. The inertia at impact, by crushing and breaking the lead cone, releases the plunger. For additional security a little oakum is placed between the percussion cap and the front diaphragm. The large number of premature explosions of this shell (see table already given) indicates that this is a dangerous device.

The Tice concussion fuze is represented in section by Fig. 11, Plate I. At the instant of starting, the inertia of the iron plunger (*a*) causes it to break three small brass supports (*b*), (two shown on drawing,) which, releasing the coiled steel wire spring (*c*), causes it to push the brass cylinder (*d*) down to the head of the fuze. While this is done, the pin (*e*) acting on the cork (*f*), holds in its position the glass bottle (*g*), filled with a sensitive fulminate and wrapped round with cotton to prevent accidental fracture. Being thus left exposed among the loose shot by the withdrawal of its protecting cylinder (*d*), the first graze of the shell causes the breaking of the glass and the explosion of the fulminate. The flash igniting the powder in the chamber (*h*), the diaphragm (*i*) is blown out and the bursting charge of the shell explodes. This fuze was used with Berney's incendiary shell with good results, seventy-three per cent. igniting precisely as desired.

The Schenkl combination fuze is represented in section by Fig. 12, Plate I. It consists essentially of a fuze plug, a fuze proper, and a fuze rotator. The hollow fuze plug (*k*), graduated on the top from 0 to 9 seconds, screws into the fuze hole and differs from ordinary plugs only in being closed at the bottom, and being pierced by a series of small holes, beginning near the middle of the plug where it enters the interior cavity of the shell, and extending to its lower end on a line curving once nearly around the entire plug from left to right. The fuze proper consists of the ordinary time fuze composition (*l*); a metallic case (*m*) containing it, which is pierced with small holes like the fuze plug, but with this difference, that they wind around the fuze case from right to left; a percussion cap (*h*) covered with highly inflammable "soluble cotton" in the space (*i*); and an iron plunger (*a*), held securely by a safety pin (*g*), (which just previous to firing is removed,) and lightly by a pin of soft metal (*b*), which is broken by the inertia of the plunger at the instant of discharge, thus allowing the plunger to explode the percussion cap and ignite the fuze composition. The fuze rotator is composed of the remaining parts shown on the drawing, which screw into the plug as indicated and hold firmly the metallic fuze case by the two dowels (*n*). It is kept in place by the latch pin (*e*) fitting any of a series of twenty holes in the top of the fuze plug, of which two are shown in the drawing. By raising the latch spring (*c*), this latch pin is lifted up, and the fuze rotator and fuze proper may be turned round until an arrow head on the side of the rotator is placed opposite any desired figure of the graduation on the top of the fuze plug.

These figures are so arranged that when the fuze composition has burned the indicated number of seconds, the hole which it has reached in the metallic fuze case shall be opposite to the corresponding hole in the fuze plug, and the flame will thus dart into the interior of the shell and ignite the bursting charge. If, however, the shell strikes any resisting object before the expiration of the time for which the fuze has been set, its rotary motion gives a rapid turn to the rotator and fuze proper, and the explosion at once takes place. A piece of tape, not shown in the drawing, is bound around the fuze rotator to hold the safety pin in position until its removal for use. This fuze is made of soft metal, which we found was liable to melt in about seven or eight seconds, and hasten the time of explosion accordingly; a defect which might be remedied by making the case of brass. It also often failed to ignite, possibly from the pin (*b*) being made too strong.

The Sawyer combination fuze is shown in section on Fig. 13, Plate I. It consists of a fuze plug and time fuze case, which are kept separate until the loading of the gun. The former (*a*) screws into the lead coating of the shell, (see Fig. 13, Plate II,) extended to the front for that purpose. This plug contains a ring of fulminate (*b*), which, in case the time fuze fails, is crushed against a projecting ring at the front end of the iron part of the shell, thus making the fuze percussion. The time fuze case consists of a brass cylinder (*c*) open at both ends, and pierced by four rows of small holes (*h*), (two rows shown on drawing) extending parallel to the axis from the rear end to the point of entrance into the interior cavity of the shell. These holes are so placed as to mark quarter-seconds upon the ordinary seven-second paper fuze (*d*) contained within the case. A gimlet exposes the fuze composition at the hole marked with the desired time of flight. This time fuze is ignited at the instant of discharge by the brass and lead plunger (*e*), held ordinarily in place by the swell in its lead base indicated on the drawing, and by the roughened interior surface of the brass case,

The inertia of the plunger at discharge drives the ring of fulminate (*i*) upon the iron ring (*g*), and by its explosion ignites the time fuze. As already stated, this time fuze case is kept separate from the shell until the time of use, being replaced by a cork to close the orifice in the fuze plug. This fuze proved to be a good one, so far as our limited practice with it afforded means of judging.

The Bormann time fuze for smooth-bore guns and howitzers is too well known to require description. It is explained in most text books on artillery.

The following table shows the comparative excellence of these different fuzes, as carefully tested in our batteries:

Service test of United States fuzes.

Kind of fuze.		Number used.	Uncertain.	Number tested.	Burned well.	Burned variably.	Did not burn.	Percentage serviceable.
Parrott	Percussion	9377	3431	5946	5059	122	765	0.85
Parrott	Time	3853	1033	2820	2143	251	426	0.75
Schenkl	Percussion	3308	1167	2141	1776	123	242	0.82
Schenkl	Combination	408	247	161	89	25	47	0.55
Dyer	Time	296	94	202	143	......	59	0.71
Absterdam	Percussion	236	48	188	99	27	62	0.53
Tice	Concussion	41	....	41	30	1	10	0.73
Sawyer	Combination	135	47	88	75	2	11	0.85
Bormann		1213	559	654	508	37	109	0.77

We had little to learn from the confederates in the matter of rifled gun fuzes, which were inferior in their service to our own. Their standards were paper time fuzes and percussion fuzes, both closely resembling our patterns.

The former were generally more nicely graduated than ours by an accurate printed scale pasted upon them, for use when cutting was required. Tufts of cotton, smeared with an inflammable compound, were often attached to the front end of the paper case by a flattened wire ring. The object, of course, was to insure ignition; and for this purpose the "McEvoy attachment" was occasionally used with rifled guns of large calibre. It consisted (Fig. 9, Plate I) of a hollow wooden cylinder (*b*) containing a friction primer (*c*), enclosed in lead (*d*), and hung on the wire (*f*). For service it was pressed upon the projecting end of the paper fuze, which was fired by the flash of its friction primer, ignited by inertia at the instant of discharge.

The confederate percussion fuzes differed from our own in having an immovable front diaphragm, (and rarely a simple wire for a rear diaphragm,) in containing unloaded plungers, and in the manner of holding the latter in place. This was usually done by a safety pin, which passed through a hole in the side of the plug, and supported the shoulder of the plunger. This pin, of course, was removed and the plug screwed firmly down into the shell preparatory to firing. One primitive pattern of fuze, acting by a friction primer (*a*) ignited by the inertia of an attached lead bullet (*b*), is shown on Fig. 14, Plate I.

The English fuzes with the Armstrong battery, the capture of which has been mentioned, were so complicated and costly as to be curiosities. They belonged to the percussion and the combination classes; the latter in two separate parts put up in different tin boxes.

The percussion fuze is shown by Fig. 15, Plate I. The lead plunger (*a*) is charged with powder in the cavity (*b*) and carries on its front end a hollow wooden cylinder (*c*) containing a sensitive fulminate (*d*). It is held in place by the safety pin (*e*) (to be removed at the moment of firing,) and by four small leaden projections resting against the lower part of the brass ring (*f*)—see Fig. 10, Plate I, which shows two of four similar projections marked (*e*). The inertia of the plunger at impact breaks off these projections and drives the fulminate forward against the steel needle (*g*), which explodes it and the powder in the cavity (*b*) and thus ignites the shell. A rubber washer (*h*) insures close contact between the fuze and shell. The fulminate is covered at top by a thin circular plate of sheet brass and a paper, and at bottom by a paper and two rings of rubber, probably to attach it to the plunger. Central holes through this rubber and a slit in the thin leaden plate supporting the cylinder admit the flame to the cavity (*b*). This cavity is closed at bottom by paper, separating it from quick-burning composition in the cavity (*i*), which is hermetically sealed at the bottom by a thin circular plate of sheet brass. The importance of keeping the fuze from exposure to dampness has been so highly estimated that the circular top (*k*) is permanently held in place by crushing the surrounding brass upon it, thus rendering it necessary to destroy the fuze in order to inspect it. It will be noticed that there are seventeen different parts in this percussion fuze, or about double the number in our patterns. It is put up singly in small tin boxes packed with oakum. This fuze is probably used only with the Armstrong shell. (Fig. 71, Plate VI.)

The combination fuze (Fig. 10, Plate I) consists of a concussion fuze and percussion fuze, which may be used together in the segment shell. Either may be used separately in it if desired. The concussion fuze may also be used separately in the shell shown on Fig. 71, Plate VI.

The method of charging the segment shell is so peculiar that Fig. 10, Plate I, has been drawn to illustrate it in its most complicated shape, *i. e.*, with the combination fuze. There are three separate pieces, the two fuzes and the bursting charge case. The bursting charge is contained in an iron case (*a*) exactly fitting the cavity of the segment shell, and closed at *each* end (to prevent mistakes in loading) with a brass cap (*b*) cemented to the iron and pierced with a hole closed by a circular piece of leather (*c*). The object of this case is to develop fully the force of the bursting charge; to expand it uniformly upon the inner surface of the shell; and to guard against accidents by keeping the powder separate from the shell until the moment of firing.

The percussion fuze is identical with that shown on Fig. 15, Plate I, except that it enters the cavity of the shell, and has a top pierced with four holes (two shown and marked (*d*) in drawing) designed to allow the flame of the concussion fuze above it to pass through and ignite its composition. This fuze is intended to guard against a failure in the concussion fuze to ignite.

The percussion fuze is of beautiful workmanship in brass, extremely complicated, and only to be fully examined by means of the saw. It consists of the fuze plunger, the index piece, and the clamping cap. These parts are readily distinguished in the drawing, from the washers at their points of contact being represented by heavy black lines. The fuze can burn only four seconds, and is accurately graduated to twentieths of a second by a band of printed paper pasted around the larger circumference of the fuze plug on the surface shown in section by the line marked (*f*). By loosening the clamping cap (*g*) a rotary motion can be given to the index piece around the central axis. The general section of this piece is shown on the left side of the drawing, but at one point it bears an index shown in mid-section on the right side. An arrow-head pointing downward is marked on the surface of this index shown by the line (*h*). By bringing this over the desired second on the graduated scale, and holding it there by tightening the clamping cap (*g*), the fuze is set for use. It is ignited in the following manner: the concussion at discharge develops the inertia of the brass ring (*i*), causes it to break the brass wire (*k*), and carry its fulminate bedded in fuze composition (*l*) upon the steel needle (*m*), also surrounded by fuze composition. The flame thus generated passes through the hollow plunger and out by the passage indicated on the drawing, terminating with the escape hole (*n*). It passes on its way the exposed end (*o*) (seen in elevation) of a ring of fuze composition seen in section at (*p*), which encircles the top of the fuze plug as in the Bormann fuze; (the end (*o*) corresponds to zero in the graduation.) This ring of composition burns until it reaches the index (*h*), where it ignites the quick composition (*r*), which in turn fires the similar ring encircling the axis, (seen in section *s*). This latter ring communicates through the duct (*t*) with the chamber (*u*).

This chamber is charged with fuze composition, which being ignited throughout its whole extent through the hole in its middle, the flame drives out the soldered circular plate of sheet brass (*v*) and passes onward to the bursting charge of the shell.

As shown in the drawing, the fulminate is enclosed in a hollow brass cylinder, which screws into the plug and is then soldered in so that it cannot be removed. The escape hole (*n*) is covered by black paper. Thus all moisture is carefully excluded. The fuze has a rubber washer (*w*) to tightly close its junction with the shell.

Without experience with these Armstrong fuzes, no opinion as to their merits can properly be formed, but they are clearly much more costly and complicated than those in our service.

RANGES.

For most of our rifled artillery, printed tables of ranges are readily to be found. For the smooth-bore bronze 6-pounder, rifled on the James system, none is known to me. Captain A. P. Rockwell, commanding the 1st Connecticut light battery, deduced an experimental table of ranges for this gun when stationed in the Department of the South, which is here given:

Ranges of 6-pounder bronze gun, rifled on James system.

Charge.	Projectile.	Elevation.	Range.	Time.
Lbs.		° ′	*Yards.*	
1.25	Schenkl shell, weighing twelve pounds.	1 00	567	......
		2 00	900	......
		3 00	1100	......
		4 00	1440	......
		5 00	1700	......
1.25	Hotchkiss shell, weighing fourteen pounds.	1 00	500	4
		2 00	800	
		3 00	1040	
		4 00	1320	
		5 00	1530	
		5 30	1730	

PRECISION OF FIRE OF RIFLED ARTILLERY.

The target practice of the 1st Connecticut artillery in the defences of Washington was accurately recorded in order to test the precision to be counted upon from rifled guns in the hands of troops. It was limited on account of the want of a good ground for safe firing with long-range guns, from which the ricochet is excessive and eccentric.

The following table exhibits the data collected. With proper pointing apparatus better results would be obtained:

Precision of fire from rifled artillery.

Gun.	Fort.	Total number of shots.	Range.	Number of shots within the following distances from centre of target.						Mean impact from centre of target.
				5 ft.	10 ft.	20 ft.	30 ft.	40 ft.	50 ft.	
			Yards.							*Feet.*
100-pdr. Parrott ..	Worth ..	35	1820	3	3	21	27	30	34	23
100-pdr. Parrott ..	Worth ..	13	2220	......	3	6	8	12	13	23
30-pdr. Parrott ...	Barnard.	9	1030	1	2	6	9	......	......	16
4. 5-inch Ordnance.	Worth ..	40	1820	2	10	24	34	38	39	19
4. 5-inch Ordnance.	Worth ..	10	2220	......	2	3	9	......	......	27
12 pdr. Whitworth	Worth ..	5	2220	......	......	3	3	5	......	25

PENETRATION OF RIFLED ARTILLERY.

In May, June, and July, 1863, the following experiments were made to determine the effect, upon earthen magazines and parapets, of our rifled siege projectiles ; and especially of our *percussion shells*, as the amount of their penetration before explosion was a matter of great doubt. The particulars as to the targets and the ammunition used will be given first; then a detailed description of the effect of each shot ; and lastly the general conclusions deduced from the firing.

Target No.1 was a bluff natural bank of sandy soil. Target No. 2 was the interior slope of a parapet eight months old constructed near battery Garesché, and having an ordinary nine-inch fascine bedded at its foot. (See Fig. 20, Plate I.) Target No. 3 was a field magazine constructed at Fort Barnard eighteen months before, of sandy soil well rammed, and supported by a framework of oak logs. (See Fig. 21, Plate I.) The magazine was sodded, but except at shots 16, 17, and 18, the surface was slightly cut away so as to present a plane perpendicular to the line of fire. This target was tested under the authority of Brigadier General Barnard, Chief of Engineers, defences of Washington. The earth of targets No. 2 and No. 3 was so solid as to leave a clean hole somewhat larger than the projectile, viz: for 30-pounder Parrott, nine inches; for 12-pounder Whitworth, six inches; for 24-pounder smooth-bore, twelve inches. No craters were made by bursting shells.

The following table describes the ammunition used and its tests:

Projectiles tested for penetration.

Gun.	Calibre.	Charge.	Projectile. Kind.	Projectile. Length.	Projectile. Weight unfilled.	Target.	Range.	No. of shots.
	In.	*Lbs.*		*In.*	*Lbs.*		*Yds.*	
100-pdr. Parrott....	6.4	10.0	Parrott, solid shot	13.5	98.5	No. 1	1800	5
							2200	1
4.5-inch Ordnance...	4.5	3.5	Schenkl, shell ...	12	27.0	No. 1	1800	9
30-pdr. Parrott	4.2	3.25	Parrott, solid shot	9	30.5	No. 3	40	3
30-pdr. Parrott	4.2	3.25	Parrott, shell....	12	27.5	No. 3	10	5
						No. 2	100	1
20-pdr. Parrott	3.67	2.0	Parrott, shell....	9	14.25	No. 2	100	2
10-pdr. Parrott, (old)	3.3	1.0	Parrott, shell....	6	10.6	No. 2	100	1
12-pdr. Whitworth..	3.0	1.75	Solid shot	9	12.0	No. 3	15	3
24-pdr. siege gun....	5.82	6.0	Round shot.......	....	24.3	No. 3	15	3

The following table gives a detailed description of the results of each shot, arranged in the order indicated in the last table :

Experiments upon penetration of rifled artillery.

Number of shot.	Projectile.	Range in yards.	Target.	Remarks.
1	100-pdr. Parrott, solid shot	1, 800	No. 1	Penetrated 10. 5 feet, then went 1 foot straight down and 2 feet to right; rested point down.
2	......do	1, 800	No. 1	Penetrated 8 feet, curving 2 feet to right; base rested nearly to the front.
3	......do	1, 800	No. 1	Penetrated 12 feet, curving 1 foot down; point rested to right and rear.
4	......do	1, 800	No. 1	Penetrated 8 feet straight; struck a hard piece of clayey sand.
5	......do	1, 800	No. 1	Failed to take groove; cut a furrow 10 feet long in loam.
6	......do	2, 200	No. 1	Struck on ground which rises about 9 degrees. From point of impact to where the shot rested was 22. 5 feet. For the first 8 feet it made a furrow, throwing up the soil. It then penetrated 14. 5 feet, straight into a loam hard enough to require the pick, and rested point to the front.
7	Schenkl 4. 5-inch shell ...	1, 800	No. 1	Penetrated 5 feet; rested point to right; did not burst.
8	......do	1, 800	No. 1	Penetrated 15 inches of slightly decayed maple stump and 1 foot of sand, and then burst.
9	......do	1, 800	No. 1	Penetrated 5 feet straight, then turned 45 degrees to right and burst. The point of the shell was found turned to the rear.
10	......do	1, 800	No. 1	Penetrated 4. 5 feet, turning 1 foot to the right. Shell rested vertically, point upward.
11	......do	1, 800	No. 1	Penetrated 5 feet straight. Burst with point upward.
12	......do	1, 800	No. 1	Penetrated 9 feet straight, and then 2 feet more, curving 45 degrees to right.
13	......do	1, 800	No. 1	Penetrated 7 feet, curving to right 1. 5 foot.
14	......do	1, 800	No. 1	Penetrated a chestnut log 6 inches thick, and then 5. 5 feet of earth; rested point to right. Drove in fuze plug, but did not burst.
15	......do	1, 800	No. 1	Penetrated 4. 5 feet straight: rested point up.
16	30-pdr. Parrott, solid shot.	40	No. 3	Penetrated 10. 5 feet straight, making hole 9 inches in diameter; sloping downward 5 degrees. Depression of gun 2. 25 degrees. Shot lodged with point to front, and turning downward about 45 degrees.
17	........do	40	No. 3	Penetrated 11 feet straight, rising 1 foot. Depression of gun 2.25 degrees. Shot lodged with point to right, and slightly turned downwards.
18	........do....	40	No. 3	Penetrated 9.5 feet nearly straight, rising 7 degrees and lodging point to right, slightly turned down. Depression of gun 2 degrees.
19	30-pdr. Parrott, shell.....	10	No. 3	Percussion fuze. Penetrated 6 feet, making a hole 9 inches in diameter, rising 5 degrees; then rose 1.5 feet in 2 feet; then burst, making a spherical cavity 1.5 feet in diameter, of which the farther side was 9 feet from point of entrance. The entire hole curved about 6 inches to the left. Depression of gun 4.75 degrees. The inside of the cavity was much blackened and cracked in every direction to a depth of 4 inches. Shell burst into about 30 fragments.

Experiments—Continued.

Number of shot	Projectile.	Range in yards.	Target.	Remarks.
20	30-pdr. Parrott shell.....	10		Percussion fuze. Depression of gun 9.5 degrees. Shell entered 6 inches below terre-plein; penetrated 3 feet of earth, then 6 inches of the oak string-piece of the magazine, and burst at the edge of the magazine excavation. One large piece made an impression of its shape upon the outside of the magazine planking. Several small pieces were found in the excavation, and two in the magazine itself. The point of the shell rested on the magazine box 5 feet from where it burst. Two quarts of dirt were blown through the ventilator; in other respects the magazine was uninjured. It is probable, but not certain, that if charged with powder it would have been exploded by this shell.
21	........do..........	10		Depression of gun 6.75 degrees. Hole rose 2 degrees. Penetrated 4 feet straight; then 2 feet, curving 30 degrees to the right; then lodged in the ventilator box, point to the rear; base of shell 6 feet from point of entrance; time fuze; the plug was crushed and small fragments chipped from the point of the shell; did not burst.
22	........do.......	10	No. 3	Did not take the grooves; depression of gun 4.75 degrees. Shell penetrated 8 feet straight, rising about 7 degrees; then 2 feet, rising 1 foot; time fuze, with loose powder placed upon its end; point of shell bruised; did not burst.
23	........do.....	10	No. 3	Percussion fuze. Depression of gun 9.5 degrees. This shell was fired at the hole made by shot No. 20. It went through the centre (9 inches) of the oak string piece of the magazine, then glancing up into the oak cross timber, 6 inches from string piece, made a furrow of the size of the shell in its lower part for 4 feet. It then burst, breaking through the magazine covering and three frames. Its effect on the magazine box was limited to about 8 feet of its length and 3 feet of its width. The point of shell and other fragments were found in the magazine, which was filled with smoke.
24	........do.......	100	No.	Shell unfilled. It passed horizontally through 4 feet of earth just below tread of banquette; then through the fascine, which deflected it upward in a gentle curve within the parapet for 5.4 feet, where it rested in a nearly vertical position. Angle of rise from fascine to point of rest 30 degrees.
25	20-pdr. Parrott, shell.....	100	No. 2	Shell unfilled. It passed through 4 feet of earth just below tread of banquette; then through the fascine, which broke it into 6 fragments, (thickness of sides of shell, 0.6 inch;) the base was then deflected downward into the natural soil 4 feet, at an angle of 25 degrees; the two other principal pieces were deflected downward into the natural soil 4.3 feet, at an angle of 35 degrees.
26	......do	100	No. 2	Shell unfilled. It passed through 4.2 feet of earth just below tread of banquette, and through the fascine, which broke it into three principal fragments. The base continued its course horizontally into the parapet 4.5 feet. The other two pieces were deflected upward for 2.5 feet, at an angle of 35 degrees.

Experiments—Continued.

Number of shot.	Projectile.	Range in yards.	Target.	Remarks.
27	10-pdr. Parrott, shell.....	100	No. 2	Shell unfilled. It struck interior slope 2 feet below crest, passed 8 feet through parapet, gradually ascending, and left it with considerable velocity.
28	12 pdr. Whitworth, solid shot.	15	No. 3	Depression of gun 3.75 degrees; depression of hole 1 degree. Penetrated 8 feet straight, then 3 feet more, curving 1.5 feet to left; lodged point forward and upward 45 degrees.
29	......do	15	No. 3	Depression of gun 1.25 degrees; ascent of hole 3 degrees. Penetrated 7 feet, curving to the left 6 inches. Rested vertically, point upward.
30	......do.......	15	No. 3	Depression of gun 1.3 degrees. Penetrated 9 feet, rested with point to left. The hole opened out from 6 inches at entrance to 1 foot diameter, and retained the latter size until within 2 feet of end.
31	24-pdr., round shot.......	13	No. 3	Depression of gun 4.25 degrees; depression of hole 7 degress. Penetrated 9 feet straight.
32	......do..........	15	No. 3	Depression of gun 2.75 degrees. Penetrated 8 feet straight and horizontally.
33	......do.......	15	No. 3	Depression of gun 2.75 degrees. Shot penetrated 8 feet straight and horizontally.

At the siege of Yorktown, an elongated projectile fired by the confederates from a rifled gun 6.4-inches calibre, struck in the centre of the exterior slope of our battery No. 1, and penetrated nine feet. It then burst, blowing out no crater but cracking the superior slope in every direction by fissures from one to two inches wide. The range was 3,900 yards.

The 7-inch Brooke rifle in the Howlett battery (Dantzler) on James river, made, at a range of 2,700 yards, deep penetrations into our parapet at Battery Spofford and in the bluff bank below it. In the latter (sandy clay) the penetration was found by digging out a projectile, to be sixteen feet. A twelve foot parapet at Fort Brady, north of James river, was repeatedly pierced (penetration about fifteen feet) at a range of 3,500 yards. The shells would lodge so near the revetment as to blow it into the battery by their explosions; and the work, (soil, clay and sand) had accordingly to be strengthened.

The following unpublished experiments were conducted in July, Angust, and September, 1863, by Lieut, Colonel J. A. Haskin, in charge of the defences north of the Potomac. The target was a butt of new earth, 383 yards distant from the guns of Battery Cameron, from which the firing was made.

Penetration of rifle projectiles into new parapets.—Colonel Haskin's experiments.

Gun.	Projectile.	Charge.	No. of—		Penetration.				Cavity after explosion.		Remarks.
			Shots.	Explosions.	Least.	Greatest.	Mean.	Before explosion.	Depth.	Diameter.	
		Lbs.			*Feet.*	*Feet.*	*Feet.*	*Feet.*	*Feet.*	*Feet.*	
100-pounder Parrott	Solid shot	10	2	......	15	18	16.5				
Do	......do	10	1	......	17	17	17	......	......	......	In new earth, 12 feet; in natural bank, 5 feet.
Do	Percussion shell	10	4	0	10	16	13.8	......	......	......	
Do	......do	10	1	1	11	11	11	8	3	2.5	Just reached natural earth.
30-pounder Parrott	Solid shot	3.5	4	......	9	12	10.8	......	......	......	
Do	......do	3.5	1	......	3	3	3	......	......	......	In natural bank; struck a stone.
Do	Percussion shell	3.5	1	1	12	12	12	9	3	2	
4.5-inch Ordnance	Hotchkiss shell, filled	3.5	1	1	6	6	6	?	......	......	Cavity not enlarged.
Do	......do	3.5	1	1	4	4	4	?	......	......	In natural bank; cavity not enlarged.
Do	Hotchkiss shell, not filled	3.5	4	......	4	6	5				
Do	Schenkl case shot	3.5	1	0	6	6	6	......	......	......	
Do	......do	3.5	1	1	4	4	4	2	2	0.8	In natural bank.
Do	Schenkl case shot, not filled	3.5	2	......	8	9	8.5	......	......	......	
Do	Schenkl shell	3.5	2	1	5	8	6.5	3	2	0.8	
20-pounder Parrott	Solid shot	2	6	......	3	10	8.2	......	......	......	
Do	Percussion shell	2	2	1	8	9	8.5	3	5	1.2	
10-pounder Parrott	Unfilled shell	1	4	......	5	8	6	......	......	......	
Do	......do	1	1	......	5	5	5	......	......	......	In new earth, 3 feet; in natural bank, 2 feet.
Do	Percussion shell	1	4	0	5	10	6.3	......	......	......	
42-pounder James	Solid shot, 81.5 pounds	8	4	......	11	14	12.8	......	......	......	
Do	......do	8	2	......	9.5	9.5	9.5	......	......	......	In natural bank.
Do	......do	8	1	......	16	16	16	......	......	......	In new earth, 6.5 feet; in natural bank, 9.5 feet.
24-pounder, smooth bore	Solid shot	6	5	......	7	8	7.8	......	......	......	
Do	......do	6	1	......	7	7	7	......	......	......	In new earth, 5 ft.; in natural bank, 2 ft.
Do	......do	6	1	......	8	8	8	......	......	......	In new earth, 6 ft.; in natural bank, 2 ft.

From the above data, and from the closely observed general effects of rifled siege artillery before Richmond, the following deductions as to penetration have been drawn:

1. The penetrations of our elongated rifle projectiles are very variable, depending upon the direction preserved by the axis of the shot. When this remains straight, they exceed those of round shot of corresponding weight by at least one-fourth, even at the shortest ranges; and this notwithstanding the far greater charges and the higher initial velocities of the latter. If the axis of the projectile turns, however, as a slight obstacle, even a fascine, may compel it to do, the penetration is greatly reduced, sometimes nearly one-half. There seems to be a tendency to curve upward after entering the earth.

2. Our percussion shells attain fully three-fourths of their entire penetration before bursting; time fuzes are very liable to be extinguished upon entering the dirt. Hence the former should be employed against magazines.

3. Earthen parapets of proper thickness cannot be seriously injured by the explosions of rifled shells of any calibre less than 6.4 inches; and not by these if the garrison is active in repairing damages.

4. With the ordinary clay loam of Virginia, parapets and magazines must have a minimum thickness of at least sixteen feet, to resist the 100-pounder projectile, (calibre 6.4 inches,) and twelve feet to resist smaller calibres ; and this, too, when well rammed and fully settled. In new earth not rammed these thicknesses must be materially increased. Hence, when practicable, magazines should be covered from direct fire by the parapet, and, where the site will permit, be sunk entirely below ground in order to economize labor. Additional security is given by a covering of fascines designed to deflect the projectile from its straight course.

GENERAL CONCLUSIONS RESPECTING RIFLED ARTILLERY.

A few general remarks may be added, based upon our experience with rifled guns accompanying an army in the field.

1. For field batteries in a wooded district like Virginia, rifles have by no means superseded the use of smooth-bore guns. In the army of the Potomac, I believe it was the opinion of the best artillery officers, that field batteries should be organized in the proportion of one rifled 10-pounder battery to two light 12-pounder batteries, (smooth-bore.) This opinion was founded upon the fact that at short ranges the latter are more effective from their larger diameter of bore, which, especially for canister and case shot, is a very important consideration. For siege artillery, however, the rifled gun has entirely superseded the smooth-bore.

2. The 20-pounder Parrott (calibre 3.67 inches) proved to be too small to give the precision of fire demanded of a siege gun, and to be too heavy for convenient use as a field gun. Moreover, its projectiles did not seem to take the grooves as well as those of either smaller or larger calibres. The gun was accordingly not regarded with favor.

3. The two siege batteries of 4.5-inch Ordnance guns, which accompanied the army of the Potomac in all its movements from Fredericksburg until the final crossing of the Rapidan, were of great use, from their superior range and accuracy, in silencing troublesome field batteries and in other field service; and could be moved with the reserve artillery without impeding the march of the army.

4. The organization of a powerful rifled siege train afloat proved of great advantage when the sudden concentration of a heavy fire upon the position selected by the confederates for their final battle became desirable. The troops were not delayed by it on the march, and it was at once available when needed. Moreover, we were thus enabled to use larger calibres than could be moved by land. For dragging the guns to the batteries, mules proved to be better than horses, and mortar wagons superior to large sling carts.

5. Without questioning the wisdom of composing each field battery of a single calibre and class of guns, as was

always done after a little experience in Virginia, we found great advantage in placing a few smaller rifled guns in heavy batteries, where barbette carriages are required. Thus battery Spofford, on James river, was composed of two 100-pounder and three 30-pounder Parrotts. It was subjected to a heavy fire of 7-inch and 8-inch Brooke rifles and 10-inch columbiads, at a range of about 2,700 yards. The enemy were so annoyed by the rapid practice of the 30-pounders at their embrasures as to make interrupted and wild firing, thus enabling the practice of the 100-pounders to be deliberate and effective. The 30-pounders played the part of sharpshooters, who can often destroy precision of fire in a field battery.

6. In placing rifled artillery in position, great errors may be made either by too much scattering or by too much concentrating the guns. In the former case due watchfulness and discipline cannot be maintained among the detached squads of cannoniers, necessarily left without the proper number of officers. In the latter case the enemy may so combine the fire of his different batteries as to silence even a superior armament. My experience led me to think that no battery should be manned by less than half a company, nor by more than one company; in other words, should not consist of less than four, nor of more than eight, or, at the most, ten guns. These remarks upon massing guns of course are not to be understood to apply to *the fire*, which should be powerfully concentrated upon the important points of the enemy's line.

7. Iron-clads can be seriously annoyed by the practice of small rifled guns well covered within proper ranges. Thus, as admitted by the Richmond papers, the sudden firing at daylight, on October 22, 1864, of an unsuspected battery of three 30-pounder and four 20-pounder Parrotts, at a range of 1,500 yards, upon the confederate fleet in Graveyard bend on James river, repeatedly penetrated the smoke-stack of the iron-plated ram Fredericksburg, wounded six of her gunners, and started one of her plates, besides dam-

aging a gun-carriage and wounding five men on the wooden gunboat Drewry. The confederate fleet promptly withdrew from the bend, and were afterwards more cautious. The heavy return fire wounded one of our men, doing no other damage.

In the attempt to pass down James river on the night of January 23–4, 1865, the confederate fleet suffered severely from the fire of our water batteries, the only serious opposition it received. The Drewry was blown up and sunk by a 100-pounder Parrott shell, fired from battery Parsons at a range of about 1,500 yards. The rest of the fleet (three iron-clad rams) retired, after being struck about seventy-five times by 100-pounder Parrotts, 30-pounder Parrotts, and 10-inch sea-coast mortars.

200 pdr. and 100 pdr. Parrott battery (No. 1) before Yorktown.

CHAPTER IV.---EARTHWORKS.

Defensive lines. Redoubts with closed gorges superior to open redoubts in the battles before Petersburg. Obstacles. Shelter from projectiles, including parapets, looped shields, mantelets, bomb-proofs, &c. Data for estimating soldiers' labor. Magazines. Boyaux.

Earthworks are so closely allied to siege artillery that this paper would hardly be complete without a few remarks upon them.

Our system of intrenchments at Petersburg consisted, in general terms, of a series of field-works, each capable of containing a battery of artillery and a strong infantry garrison. These works were closed at the gorge, were protected with abatis and palisading, were often supplied with bomb-proofs, and were located, at intervals of about 600 yards, on such ground as to well sweep the line in front with artillery fire. They were connected by strong, continuous infantry parapets, with obstacles in front. The main Bermuda Hundreds line and that north of James river were similarly constructed, as were also the defences of Washington.

The confederate system differed from this, chiefly, in having its redoubts open at the gorge, and not arranged for independent defence.

The merits of these different systems of lines have been much discussed by military writers. The Petersburg battles of March 25th and April 2d, 1865, illustrate the superiority of the arrangement adopted by our own engineers. The circumstances attending the first named battle were the following:

Hare Hill was situated near the right of our Petersburg line, about a mile from the Appomattox river. It was protected by Fort Steadman, with battery No. 10 on its right and batteries 11 and 12 on its left. The next work closed at the gorge, on the side of the Appomattox river, was battery No. 9, situated near the foot of the hill. The next work on the left of Hare Hill and its collection of batteries was Fort Haskell, situated on another hill, with a small creek between.

Fort Steadman was one of the weakest and most ill-conditioned works of the line, being unprotected by abatis in rear, being masked on its right (just in rear of battery No. 10) by numerous bomb-proofs rendered necessary by the terrible fire which habitually prevailed in this vicinity, and being only about 200 yards distant from the enemy's main line. The parapet had settled during the winter, and, in fine, the work was not well prepared to resist sudden assault.

At about 4 o'clock a. m. of March 25, three divisions of the enemy, under General Gordon, made a sudden and well-arranged attack upon the defences of Hare Hill. It was a complete surprise, and was successful. Their columns simultaneously swept over the parapet between Steadman and battery 9, over battery 10 and over battery 11, joined in rear of the fort, and carried it almost without opposition. From that time to daylight a scattering hand-to-hand fight raged among the bomb-proofs and on the flanks of the enemy's position. He assaulted Fort Haskell again and again, but failed to carry it, or battery No. 9. As soon as the light would admit, all the siege artillery from batteries 4, 5, 8, 9, and Fort Haskell, and all the light artillery which General Tidball, chief of artillery 9th Corps, could concentrate upon the position, opened and maintained a heavy fire upon the enemy. No re-enforcements could join him from his own line, owing to this fire which swept his communications; his captured position was entailing deadly loss; our reserves were rapidly assembling, and finally, about 8 a. m., they made a charge, which resulted in the recovery of our works, of all our artillery, (even including the Coehorn mortars,) and in the capture of over 1,800 prisoners. The following extracts from the confederate papers show the effects of our artillery fire:

"It was found that the enclosed works in the rear, commanding the enemy's main line, could only be taken at a great sacrifice." "The enemy massed his artillery so heavily in the neighboring forts, and was enabled to form such a terrible enfilading fire upon our ranks, that it was deemed best to withdraw."

"The enemy enfiladed us from right and left in the captured works to such an extent that we could no longer hold them without the loss of many men," &c., &c.

If the enclosed works on the right and left had not fixed a limit beyond which the enemy found it impossible to extend his lines, his great local superiority in numbers might have given us trouble.

On April 2 we, in turn, were the assaulting party. As in the battle of March 25, the sudden unexpected attack at early dawn was successful, near Fort Mahone; and the confederate redoubts being all open at the gorge and hence untenable against a flank and rear attack in force, were soon occupied along such an extent of their line as to prevent that terrible concentration of artillery fire which would have been made upon a contracted position. We succeeded all day in holding our own against a series of counter assaults, which clearly showed that the ruin then impending over the confederacy did not impair the personal courage of its defenders.

The great practical difficulty in defending lines in close proximity to the enemy lies in *keeping the troops always ready to receive assault.* The engineering system adopted on the Bermuda Hundreds lines, with this object in view, proved excellent; viz., to place from 300 to 400 yards in advance of the main line a series of small redoubts, well protected by obstructions, provided with three or four smooth-bore field guns, and garrisoned by a full artillery company armed with muskets and supported by a small force of infantry. These works could not be neglected by an assaulting column, and their vigorous defence, even if unsuccessful, would give the necessary time for manning the main lines; which the pickets, unsupported, could not in such circumstances be trusted to secure. The confederate assault of June 2 was repulsed by one of these advanced redoubts, laid out, it should be added, entirely without local flanking arrangements.

When a large isolated earthwork, with a small garrison, is to endure a siege, as at Fort Wagner for instance, it is doubtless necessary to provide flanking arrangements for the ditches; but the experience in Virginia showed that with small earthworks on the lines of a great army, the interior space thus necessarily sacrificed is of far more importance to the defence. If the line offered strong salients which crossed fire in front of each other, the close flanking of the ditches proved to be of little importance.

Regular siege approaches upon a long line of well-manned earthworks, which cannot be enveloped, and which are vigorously defended by forces nearly equal to our own, provided with smooth-bore field artillery, numerous mortars, and modern rifle muskets, we found to be so difficult as to warrant little anticipation of victory.

OBSTACLES.

Efficient exterior obstacles are of the greatest importance, but our experience showed that they should not be placed close to the ditch. The nearest line should be about fifty yards in front ; and, if possible, a second at one hundred yards distance should be added. Abatis at these distances properly pinned down, with a few telegraph wires twisted around stumps or stakes about a foot above the ground, is almost impassable, and cannot be destroyed with artillery fire, unless by enfilade. A "slashing" in forest land is very effective, but in dry weather it soon becomes liable to be burned by shells; a fate which often befell our defences of this character. Chevaux-de-frise were not used by us, but were much depended upon by the confederates, who connected them by chains at the ends. In the assault upon their works at Petersburg, on April 2, 1865, this species of obstruction but very little retarded our troops.

The following statement of the obstructions on the exposed portions of the lines in front of Bermuda shows the importance attached to this means of defence by the confederate engineers: 1st, a fraised ditch to the main line; 2d, an abatis at about 25 yards; 3d, a palisading at about 50 yards; 4th, a chevaux-de-frise at about 75 yards; 5th, an abatis at about 150 yards; 6th, an intrenched skirmish line at about 500 yards; 7th, an abatis just in front of the latter. We never defended our lines so strongly, usually depending upon a slashing, or upon one or two lines of abatis or palisading, interlaced when practicable with telegraph wire. Such indeed, was the usual confederate system.

The confederates also paid considerable attention to land torpedoes as obstacles against assault, while we neglected them. At Fort Fisher, North Carolina, the elaborate system of electrical torpedoes is well known. Along and in front of the abatis of the works north of James river, near Fort Harrison, immense numbers of loaded shell, with a very sensitive fuze so arranged as to ignite when trodden on, were buried. Earlier in the war, at the siege of Yorktown, these loaded shells were also used in great numbers, and not always according to the rules of civilized warfare. At Charleston and Mobile, also, they were planted in large numbers.

SHELTER FROM PROJECTILES.

In ordinary soils, parapets likely to receive a heavy fire from field artillery should be not less than 12 feet thick, of well-rammed earth; to resist the fire of modern siege guns this must be increased to 16 feet; to resist the 7-inch and 8-inch rifled sea-coast guns, not less than 20 feet will suffice. All these dimensions must be increased when, as is generally the case in the field, ramming is not attempted, and the fire is expected to begin before the earth has had time to settle. In quartz sand a less thickness will suffice.

In sandy clay, like the soil in Virginia, the width of the berme should be at least half the depth of the ditch, to resist the crumbling action of frost. The ultimate practice in the defences of Washington was to prolong the exterior slope to the bottom of the ditch.

Revetment for interior slopes in Virginia was usually

9

made of logs or poles; rarely of gabions, fascines, hurdles, or sand bags.

The parapet of field-works on the lines of an army should not be made more than eight feet high. The curvature of the trajectory, both for artillery and musketry, is so great that almost the only utility of defiladement is to conceal the interior of the work from the view of the enemy. This object is more readily attained by a few sand bags, or by a hurdle screen looped for musketry, than by attempting the construction of enormous parapets with soldiers' labor,* which, unless an immediate attack is apprehended, is far less than that of paid workmen.

The great accuracy of the rifled musket rendered it necessary at several points of the lines to construct covers for

* To determine what would be a safe estimate for the labor of good troops, some careful notes were taken when fortifying the siege-train depot at Broadway Landing. The soil was sandy clay; the weather excellent; the troops, a detachment from the 15th New York volunteer engineers, company M, 3d Pennsylvania artillery, and a few men of the 1st Connecticut artillery. Three thousand days' work, of ten hours each, were required; of which twenty-five hundred were spent in digging, and five hundred in cutting and placing 420 yards of pole revetment well anchored with wire, in cutting and planting 600 yards of good abatis, and in slashing several acres of timber in the surrounding ravines. The men were placed half in the ditch, (picks and shovels equally distributed,) and half on the parapet, which was seven feet high; the ditch was twelve feet wide and six feet deep. The men averaged 6.4 cubic yards of dirt moved to the day's work. Hence, with moderate profiles, in ordinary soil where the pick is needed, and the abatis, &c., easy of access, the daily amount of excavation by the detail, less one-sixth for the party placing the corresponding revetment, abatis, &c., may be estimated by allowing 1.6 cubic yards per man;—where the pick is not needed one-fourth more will be done;—that is, a detail of six hundred men will excavate and place in the parapet $500 \times 1.6 = 800$ cubic yards in the former, and $500 \times 2 = 1,000$ cubic yards in the latter case per day. My experience as an engineer in the earlier campaigns of the war accords with this rule when our best class of troops are engaged and *not expecting immediate battle.* In the latter case, their exertions are nearly doubled. In De Rouvre's "Aide Memoire d'Etat Major" the day's work is laid down at 2.6, 5.2, and 6.5 cubic yards in hard, medium, and light soil respectively. Professor Mahan, in his "Field Fortifications," gives 6 cubic yards for ordinary, and 8 to 10 cubic yards for task work with soldiers. In Laisné's "Aide Memoire du Genie" these classes of work are estimated at 10.5 and 15.4 cubic yards respectively, when laborers are employed. The latter estimate accords well with that of experienced contractors upon the levees of the Mississippi river.

the head in firing, reconnoitring the enemy's line, &c. We usually made them of sand bags, and the confederates of logs roughly looped at three feet intervals, but both devices were occasionally employed by each party. To protect the sharpshooters something more was necessary, and a log having a square loophole of minimum size, strengthened by a small musket-proof iron plate on its exterior was adopted. These plates often bore the marks of having been repeatedly struck, so deadly was the precision of this description of fire.

To protect our artillery, mantelets were often required. They were furnished by the Engineer Department—two patterns of combined wood and iron, and one of rope. Both of the former met with little favor, owing to the impossibility of modifying the openings when used in oblique embrasures; owing to the shattering caused by the blast of short guns; and owing to the dangerous splinters thrown off when struck by artillery. The latter—dimensions 5 by 4.5 by 0.5 feet, with opening 1.6 by 1.3 feet—was too heavy for convenience, weighing over five hundred pounds. The penetration into it of a rifled musket ball fired at twenty yards was 2.5 inches. A later pattern, only four inches thick, served every purpose, although occasionally balls striking between the ropes would pass through. It was made of four thicknesses of three-inch rope, each of the two outer layers being in one piece bent vertically, and the two inner of short pieces laid horizontally, the whole well tied together.

The chief object of the mantelet is to conceal the cannoniers from the enemy's sharpshooters by a screen *supposed* to be bullet proof. When this is done but few shots are directed at it, and if they occasionally penetrate, the risk must be incurred rather than allow excessive weight.

Our mantelets were hung on horizontal poles supported upon forked uprights, the elasticity of which greatly increased the resistance; thus, on one occasion when one of the later (four-inch) pattern in battery No. 9 on the

Petersburg lines was struck by a ten-pounder rifled shot, fired at six hundred yards, the shot was deflected after breaking the lashings and throwing down the pole supports, and was so much checked in velocity as to then knock a man down without killing him. The confederates made some use of small ring mantelets placed upon the gun itself and pierced with a slit to admit of pointing. We tried this plan ourselves at the siege of Yorktown; the gun, of course, was run back and loaded under cover of the parapet; the device seemed a good one.

The confederate batteries on Chesterfield Heights were interesting studies, from the ingenuity displayed in obtaining shelter. They had long annoyed the right of our line before Petersburg with an enfilading fire, and had received much attention from our batteries in consequence. The range was about one thousand six hundred yards, and to avoid the effects of our shells, which often exploded in the embrasures, an extraordinarily long and narrow shape was given them. The most troublesome battery (4.2 inch siege rifles) was half sunken, and had a parapet thirty-three feet thick, with embrasures which were hurdle-revetted two feet wide at the throat, and only four feet wide at the exterior crest! Substantial traverses with good splinter proofs were made on each side of the guns in this battery, and solid bomb-proof quarters were near by. In locating one of these batteries for Whitworth guns, much skill had been displayed in lessening the effect of our fire by stratagem. The bluff of the Appomattox is here level at top; the battery was placed about four hundred yards from the edge, with a ploughed field in front and a skirt of bushes on either side, so that from our lower bank we could obtain no flank view. An error had clearly been made in under-estimating the distance of the battery from the edge, and our shells throwing up dust from the field in front had been supposed to be striking the parapet. With horizontal fire this was not a perfect protection, as the ricochets would annoy the gunners, but for effective mortar fire it was fatal.

Many plans were tested by both armies for obtaining protection against mortar fire. Bomb-proofs of every conceivable shape, generally made in half excavation of log framework covered by earth, answered this purpose for men off duty. Each soldier generally made his own shelter. They were placed near—sometimes under—the parapet, and often encroached upon the space required for using the musket. In fact, the lines in some places became involved labyrinths, nearly impassable at night to one not very familiar with the locality. At the approach of winter, chimneys of sticks, or, better still, of gabions plastered with mud, were added; and the edifices assumed an appearance only to be represented by the photograph.

For the protection of men on duty, the usual plan was to place at intervals of about fifty yards cross-pieces of timber or railroad iron from the parapet to some bomb-proof or magazine just in rear, and to cover the top with earth, (one layer of railroad iron and three feet of earth sufficed.) These covers were about six feet wide at the crest line; and the sentinels could step for a moment under them to avoid descending fragments without relaxing their vigilance.

Mortar battery number 12 on the Petersburg lines, which frequently received a heavy return fire, was constructed upon an excellent plan by Captain Pride, 1st Connecticut artillery, commanding it. He placed his mortars by twos, separated by substantial traverses; at each side of which a good shelter like those just described was placed. The cannoniers could readily pass in and out to avoid fragments, and could keep all their ammunition under cover until required for use. In the end of one traverse he arranged a covered lookout, where the effect of the shots could be safely observed. Although this battery was subjected to a very heavy fire, which tore up the platforms and marked the mortars with fragments, it could not be silenced and very few casualties in it occurred.

For protection against heavy horizontal fire, the confederates adopted the following plan on the Bermuda Hundreds lines: a continuous splinter-proof was made under the parapet behind the breast-height revetment, top two feet above and bottom one foot below the banquette tread, width four feet; this was made of a heavy framework of logs which consisted of two parallel rows of uprights (uprights in each row three feet apart) supporting cap sills and roofed by logs sloped to the front and projecting over the scarp, thus forming a fraise. The shelter was well drained, and furnished cover for one man to every six feet of parapet.

MAGAZINES.

Our experience indicated that for siege guns a very simple plan, involving nothing but what could be obtained upon the spot, answered all necessary purposes for magazines in upland soil like that of Virginia. We took great care to shelter them as much as possible by the parapet, and made the chamber entirely below ground, roofed by heavy logs, (sometimes resting on plates supported by uprights, and sometimes on horizontal logs notched into each other like the sides of a log house,) and covered by earth about six feet thick. The entrance was straight and from the rear. Boards were seldom used either for the sides or the floor, which was made to drain into a hole (a barrel if practicable,) sunk near the entrance. The usual dimensions in the clear were six feet wide by five or six feet deep, length to vary according to capacity required. In no instance was one of them blown up, although often hit by shells; and very little loss of ammunition occurred from dampness even in heavy rains, such as that of August fifteenth at Petersburg when several soldiers in the low bottom were washed away and drowned.

Where so much magazine room as is required by modern rifled siege and field artillery is needed, it is not often practicable to construct magazines upon the elaborate plans laid down in the text-books. For the heavy James river water batteries, however, this was deemed necessary by both armies; especially by the confederates, whose magazines and bomb-proofs for this purpose were extremely substantial and well made.

BOYAUX.

Our experience with covered approaches indicated that absolute protection must be sacrificed to facility in turning the corners with wagons, and especially with siege guns; the latter were several times overturned in passing the angles of the boyaux leading to Fort Sedgwick, on one occasion crushing an officer to death.

APPENDIX A.---SIEGE TRAIN OF 1862.

1.—*Report of Colonel R. O. Tyler, commanding.*

HEADQUARTERS FIRST CONN. ARTILLERY,
(SIEGE TRAIN,) HARRISON'S LANDING, VA.,
July 15, 1862.

GENERAL: From the disembarcation of my regiment at Cheeseman's landing, on the 12th of April, to this date, most of the service of the regiment as heavy artillery, and otherwise, has been performed directly or indirectly under your command; at Yorktown as director of the siege, and at Hanover Court House, Gaines's Mill, Chickahominy, and Malvern Hill, the portion of the regiment engaged has been under your immediate orders.

Under these circumstances I would respectfully request here to submit a report of the services of my regiment upon the Peninsula.

You are well aware that at Yorktown, under all the difficulties of transportation, my regiment kept pace with the engineers in laying platforms and placing guns and material in position. Batteries 1, 2, 4, 6, 11, and 14, mounting forty-eight guns and thirty mortars, were fully armed and equipped, and twenty-three guns and mortars disembarked.

It is unnecessary to call attention to the amount of labor required to do this duty, for such batteries as No. 1, mounting five 100-pounders and two 200-pounder guns, and No. 4, mounting ten 13-inch sea-coast mortars—both of which exceeded in weight of metal any batteries hitherto equipped; or to say how much the heavy firing from No. 1 for four consecutive days may have had to do with the evacuation of that place. On the day of the evacuation there were six batteries of forty-eight mortars and guns ready to throw one hundred and seventy-five tons of metal daily into Yorktown. I wish to state that my regiment removed all this material from these batteries and re-embarked it.

The greater portion of this work having been completed, I reported with ten companies of my regiment to you at White House, on the 28th of May. * * * *

Upon June 20, I was ordered to bring up a battery of five $4\frac{1}{2}$-inch Rodmans and one of five 30-pounder Parrotts, and to place them in position near New Bridge. The disembarcation of the guns and material at White House commenced on the 21st of June; and upon the 24th of June these guns were in position, complete with ammunition and material, in batteries Sykes and Porter, under the command of Major Kellogg, and in charge of companies B, D, and F, commanded by Captains Dow, Cook, and Brooker.

On June 25, under your direction, these batteries opened with good effect upon the rebel batteries on the opposite side of the Chickahominy ; doing, as reported by the signal officer, much damage, dismounting the enemy's heaviest gun and compelling them to remove their camps. Upon the 26th the batteries again opened. At 6 p. m. of that day they were moved across the Chickahominy, where they reported to General Smith. Here they were joined by two 10-pounder Whitworth guns, under command of Lieutenant Sedgwick, which had been brought around, with their material, by way of Baltimore store and Bottom's bridge. Upon the following day (June 27) these batteries were placed in position upon Golding's Hill, where they were fought during the day under a severe fire.

When the services of the guns could no longer be useful the companies were formed by Major Kellogg and led into the line of infantry defending that position, a fact specially noticed by General Smith in his report.

The casualties of this day were two men killed and one wounded. The pieces were brought off by hand, and I wish specially to call your attention to Lieutenant Sedgwick, of company I, in charge of the Whitworths; which, with only twenty men, he removed by hand a distance of two and a half miles, the second gun being brought away at a time when our most advanced pickets were retiring past it. Upon the night of the 27th and the morning of the 28th the guns under command of Major Kellogg were successfully retired behind White Oak swamp; where they joined the remainder of the siege train, which had been in position and in depot at Fair Oak station in front of Sumner's corps, under command of Major Hemingway, under whose orders were company A, Captain Gilbert, company E, Captain Rockwood, company H, Captain Hubbard, and company K, Captain Ager.

The successful removing of these guns and stores from Orchard station is greatly due to the exertions of these officers, more especially of Captain Hubbard, who was left in charge of the rearmost trains.

Upon June 25 the regiment was ordered to report to General Barry for duty as heavy artillery. The companies, except those in battery at Gaines's Mill and those employed in disembarking ordnance and ammunition at the White House, were marched to Orchard station by Lieutenant Colonel White. The rapid advance of the enemy prevented the removal of my hospital from New Coal Harbor, where many of my sick and hospital attendants, including two officers, Lieutenants Faxon and Harwood, were captured.

On the 30th of June I received an order to report to you with such guns as there was ammunition remaining for. During the night of June 30 five Rodmans, five 30-pounder Parrotts, two 8-inch howitzers, and two 10-pounder Whitworths, were transported from our camp at Turkey Bend by companies F, Captain Dow ; D, Captain Cook ; B, Captain Brooker ; K, Captain Ager, and I, Lieutenant Hatfield, under Majors Hemingway and Kellogg; and, under difficulties which you will understand, were taken up the steep ascent of Malvern Hill, with their ammunition and material, the companies working all night after their previous tedious marches. The guns occupied the highest ground of Malvern Hill, and were served under your personal orders, and caused much destruction to the enemy's advancing columns. The casualties of this day were one killed and three wounded.

The companies, after working all the night of the 30th to place these guns in position, and fighting them during the whole day on the 1st of July, spent that night in retiring the siege train to the present depot near Westover Landing ;

the guns, the ammunition of which had been expended, were also retired to Harrison's bar, under Lieutenant Colonel White. I would respectfully call your attention to the fact that all the ammunition used at Malvern Hill had been transported by way of Gaines's Mill, Savage's station, and White Oak swamp to that place; and that the officers and men with the guns had been almost constantly laboring day and night from the 22d of June; and to the fact that out of twenty-six heavy guns twenty-five arrived safely at their destination.

This was accomplished under almost unheard of difficulties with mule teams constantly breaking down, driven by frightened citizen teamsters who deserted whenever the fire became heavy ; frequently teams had to be pressed into the service to replace those which had been exhausted by the labor of drawing the guns, and sometimes for miles the guns were drawn by hand by the different companies of the regiment.

One howitzer was abandoned at Savage's station, the carriage having become so disabled that it was impossible to renew it ; under direction of Lieutenant Jackson the carriage was burned and the howitzer rendered perfectly unserviceable.

To the field officers and company commanders, already mentioned, I feel that a debt of gratitude is due for the labors they performed and the difficulties they cheerfully overcame.

I wish, in addition, to mention the services of Quartermaster Robins, and Lieutenants Whittlesey and Jackson, acting ordnance officers, to whose energy the safe bringing through of the ammunition trains is mainly due, and to bear testimony to the great assistance rendered to me on all occasions by Lieutenant Pratt, my adjutant.

I am, general, very respectfully, your obedient servant,

R. O. TYLER,

Colonel First Connecticut Artillery.

Brig. Gen. FITZ JOHN PORTER,

Commanding Fifth Army Corps,

Army of the Potomac.

2.—*Report of Major Alex.Doull, ordnance officer for train of* 1862.

NEAR YORKTOWN, VIRGINIA,
May —, 1862.

COLONEL: The siege of Yorktown being terminated by the evacuation of that place by the rebel forces just when nearly the whole of the siege batteries were ready to open fire, I have the honor to submit for your consideration the following report of the work which has been performed by the officers and men of your regiment in arming the siege batteries at that place.

During the seven days that elapsed from the 26th of April to the evacuation of Yorktown, all the batteries have been fired at more or less continuously; and though the regiment had never been under fire, and is, like the rest of this army, composed of troops who have not been twelve months in the service, and who would therefore be considered in any regular artillery in the world merely as recruits, and the officers have not had the advantage of that scientific military training which is usually considered necessary for this branch of military service, and although a large part of the material employed has been of a weight hitherto unknown in sieges, and has therefore necessitated the employment of carriages and platforms usually confined to permanent works on account of the labor, care, and accuracy required in their construction, yet the condition of the batteries, and the accuracy with which all the platforms have been laid and the magazines arranged, give no indications whatever of these disadvantages.

The siege train at present in battery and under my charge consists of—

Two 200-pounder Parrotts, five 100-pounder Parrotts, at battery No. 1 manned by battery B, commanded by Major E. S. Kellogg.

Five 4½-inch guns, five 30-pounder Parrotts, at battery No. 2, manned by batteries A and H, commanded by Major L. G. Hemingway.

Ten 13-inch sea service mortars, (1861,) at battery No. 4, manned by batteries F and G, commanded by Major E. S. Kellogg.

Six 10-inch sea service mortars, (1841,) at battery No. 6, manned by battery C, commanded by Captain R. S. Burbank.

Ten 10-inch siege mortars, at battery No. 9, manned by batteries D and E, commanded by Major Thomas S. Trumbull.

Five 4½-inch rifles, at battery No. 10, manned by battery I, commanded by Captain S. H. Perkins.

In addition there are in depot at Fort Grafton four 20-pounder Parrotts, three 8-inch howitzers, four 10-pounder Whitworths, and two 8-inch mortars, (siege.)

And there have been transported to the depot and afterward given out to Colonel Warren, 5th New York volunteers, five 10-inch siege mortars and five 8-inch siege mortars, making a total of thirty rifled guns, three howitzers, and thirty-eight mortars, or seventy-one pieces of artillery; of which sixty-five were re-embarked by the 1st Connecticut artillery, and the six 10-inch sea service mortars by the 5th regiment New York volunteers.

For the manœuvring and transportation of the material there have been furnished: one gin, complete; three mortar wagons; two large sling-carts, and three hand sling-carts. In addition, one treble, two double, and one single block were obtained from Lieutenant Baylor, Fortress Monroe. With these and with temporary derricks, constructed from such timber as could be procured, the disembarking, the loading of the mortar wagons, and the mounting of the 4½-inch and 30-pounder Parrotts at Cheeseman's landing have been performed.

The mounting of the 200-pounder and 100-pounder Parrotts, and the 13-inch sea service mortars, (1861,) was done with the gin. In mounting the latter a treble block broke. This and the slipping of the block of a gin tackle while moving one of the 13-inch mortars on rollers, were the only accidents that happened during the whole of the embarking, disembarking and mounting of this armament, although a very large amount of the duty was performed at night.

The guns and mortars were hauled from the landing by the horses of the reserve artillery.

To the zeal and energy displayed by the lieutenants who have been detailed for that service, in the performance of which, as light artillery officers, they cannot be expected to take much interest, and which was especially severe on horses and harnesses, the service is much indebted.

The 100-pounder Parrotts were transported upon the

large sling-carts. This was attended with severe difficulty, even when the roads were moderately good, and I beg to suggest that advantage would be found in widening the tires of the sling-carts so as to give a larger bearing surface.

The hand sling-carts were found very useful. In disembarking the mortars and mortar beds, the pole of one of them broke at the junction of the braces and the pole. It has been repaired by the regimental artificers in a manner which appears to be an improvement. The front bolts for connecting the upper and lower straps and the bolt for connecting the braces cross through the pole about two inches apart. These are dispensed with, and the braces and straps are shrunk on the pole by an iron ring.

The 200-pounder Parrotts were transported to the battery by water, at night, and landed by being rolled on skidding.

The 13-inch mortars and the mortar shell were run into Wormley creek at night in a barge, taken up to the battery and landed at a wharf constructed for that purpose at one end of the battery.

This was done by raising the mortars by the gin, placing rollers under them, rolling them to a port cut in the side of the barge, raising them by successive lifts with the sling-cart screw worked by drag-ropes until hung as high as possible, and then running them nearly on to the platform; when they were again placed upon rollers, rolled into position, lifted by the gin, and the carriage placed underneath.

The heaviest pieces placed in position in the trenches before Sevastopol by the English were the 68-pounder gun of 10,640 pounds, and the 13-inch sea service mortar of 11,300 pounds; and by the French, the cannon de 50 of 10,190 pounds, and the mortier de 32c of 9,615 pounds. The 200-pounder Parrott weighs 16,570, and the 13-inch sea service mortar (1861) 17,120 pounds.

The guns placed in position before Yorktown exceed, therefore, in weight by fifty per cent. any guns that have ever before been placed in siege batteries. I have therefore been thus particular in detailing the manner in which they have been placed in position.

I do not believe that any very great difficulty would be found in moving the 200-pounders by the sling carts over very good roads ; but the dimensions of the 13-inch sea service mortars (forty-three inches diameter) would render it impossible to transport them without a special carriage, as the mortar when slung on the sling cart hangs twenty inches below the tires of the wheels. In the battery, ways of logs with plank spiked on top were constructed, and the ground excavated between to permit the passage of the mortar, but this could only be done upon hard ground.

In the arranging and manœuvring of the guns, the following points in which improvements might be made suggest themselves, and are respectfully submitted for your consideration:

1. That the construction of the pole of the hand sling cart be altered as above mentioned.

2. That the simplicity of the siege platform be still further increased by making no distinction between hurters, sleepers, front and rear pieces, and deck plank ; *i. e.*, by having dowel holes in all the pieces, sending the dowels in bundles to be placed in where necessary, and leaving the bolt holes to be bored in the battery.

3. That a certain supply of tools, most of them now found in battery wagons, be supplied with the gun and mortar implements at the rate of one for each five pieces, viz: one two feet rule, one auger, a mason's level, (the gunner's level not being convenient for laying platforms,) a handsaw, one cold chisel, and a battery lantern; that a magazine be allowed for every five guns, and a complete set of magazine implements, (Ordnance Manual, p. 367,) including copper adzes, and dark lanterns, be supplied for each magazine. Battery wagon D is admirably suited for depot purposes, but if each battery and battery magazine is supplied with these stores, the battery wagons are soon exhausted of them without the other stores which they contain being called for at all.

4. That a full complement of the machines and ropes specified in Ordnance Manual, page 367, be sent with each train; and that, in addition, 3-inch planking, skidding, blocking, spare tackle blocks, and a number of small tackles, consisting of wooden double and single blocks, with $2\frac{3}{4}$-inch falls, be provided.

5. In manœuvering the 13-inch mortars it became evident that the iron-shod handspikes provided were not sufficiently strong.

6. The steps of the wrought-iron mortar carriages require an upright brace or the upper one becomes crushed in mounting the mortars.

Only one battery, No. 1, of five 100-pounders and one 200-pounder Parrott, opened fire during the siege. This battery opened fire on the 1st of May, and at once drove all the rebel shipping from the wharves at Yorktown. In all, one hundred and thirty-seven rounds from the 100-pounders, and four rounds from the 200-pounder were fired. An inspection of the batteries on the river front of the Yorktown fortifications, will afford the best indication of the very accurate practice made from this battery, but the impossibility of observing with equal accuracy the results of firing at such long ranges, (3,800 to 4,700 yards,) renders it extremely questionable whether any useful results follow a large expenditure of ammunition at distances over 2,500 yards.

The following points of importance appear to have been established by this practice:

1. That the wrought-iron carriages are too rigid to resist long continued firing at high angles, the rivets cutting off completely. It is very much to be regretted that a few rounds were not fired from the 13-inch mortars to test their carriages, as it appears probable they would fail more completely in this respect.*

2. That the mode of supporting the pintle and traverse circle used in this battery is not sufficiently strong. The

*No difficulty of this nature occurred in the train of 1864–'65.—H. L. A.

object of the platform is to so connect these two parts that no relative motion can take place. The intermediate pieces should therefore act more directly than in these platforms. The pintle plates should have a larger bearing surface upon the pintle and should be strengthened by ribs. Probably the best plan of all would be to adopt a low carriage similar to that sent by Captain Parrott for the last 200-pounder, with four trucks, the use of the platform being given up by the position of the trucks; and to dispense with the pintle altogether, using rails and hollow trucks, (Report of Major Mordecai on recent English construction.) The wooden platform would then consist of two traverse circles with wooden braces and iron tie bolts.

For the service of these guns it was necessary to convey 17,047 projectiles, weighing in all 857,417 pounds. The transportation of these projectiles and of the powder and small stores into depot at Fort Grafton commenced on the 12th of April and continued until the evacuation of Yorktown on the morning of the 4th of May. During that time 726 loads were conveyed from Cheeseman's Landing to the depot, of which 527 loads were projectiles, 70 of powder, 88 of platforms, and 41 of small stores. This does not include the projectiles for 13-inch mortars, which were conveyed to the battery by water.

The weather during the siege was very varied, and the roads consequently were sometimes in so good order that the wagons made two and even three trips a day, and sometimes so bad that they were a day and a half on the road.

On Wednesday, the 16th, sixty-two wagons brought up 1,882 10-inch mortar shells; forty wagons carrying thirty shells each, and twenty-two carrying thirty-one shells, weighing 2,700 pounds. This was the maximum. On the 27th of April seventeen wagons brought only one hundred and twenty-five barrels of powder, being only seven hundred pounds and six hundred pounds to a wagon. This was the minimum. The average taken from the projectiles was 1,456 pounds to a wagon.

The estimate of transportation is exclusive of that between the depot and the batteries, for which purpose twenty-five wagons were kept constantly employed from the 21st of April. In all six hundred and thirteen wagon loads were conveyed.

In the three weeks during which these siege operations were conducted, your regiment worked, with very little relief, night and day. As soon as any battery was completed the companies to which it was assigned moved into camp near it, constructing such shelter from the enemy's fire as they could, and remained with their guns ; differing in this respect from all other troops employed in the trenches, who returned to camp out of fire as soon as their duty was finished.

When it is considered that the first siege train placed in battery by the English before Sevastopol consisted of seventy-two pieces; that they marched into Balaklava on the 25th of September, and opened fire on the 17th of October, twenty-three days afterward, and that they employed to accomplish this end all the resources of a powerful navy, and a large regular artillery skilled by constant practice in the large maritime fortresses of Gibraltar, Malta, and Corfu, in all the manœuvres of heavy artillery, and that only the same time (twenty-three days) was occupied (from the 12th of April to the 3d of May) in placing seventy-one guns in battery, many of them much exceeding in weight any that have before been used in a siege, it is evident that the labors of the 1st regiment Connecticut artillery will compare favorably with anything of the kind that has been done before.

I have the honor to be, very respectfully, your obedient servant,

ALEXANDER DOULL,

Major 2d N. Y. Artillery, and Ordnance Officer to Siege Train, 1st Connecticut Artillery.

Col. ROBERT O. TYLER,

Commanding 1st Connecticut Artillery.

APPENDIX B.---SIEGE GUNS WITH THE ARMY ON THE MARCH.

Report of Captain F. A. Pratt, commanding battery M, 1st Connecticut artillery, upon the equipment and services of the two batteries, 1862–'64.

HARTFORD, *June* 24, 1867.

GENERAL: I find, after repeated deferring, that business engagements keep me from gathering and giving as full and detailed a report of the siege batteries in the field as I intended; I will, however, give a rough report and endeavor to convey the essentials, which you can put into the necessary form for your purpose.

EQUIPMENT OF THE BATTERIES.

The equipments of the two batteries were the same. On the march they were usually together, and the history of one is nearly that of the other.

Captain Brooker and myself, with companies B and M, were detached from the regiment at Alexandria, December 5, 1862, attached to the army of the Potomac near Falmouth, Virginia, and placed in command* of two four-gun batteries of $4\frac{1}{2}$-inch rifles. With these we were in action at the battle of Fredericksburg, beginning December 11, 1862. We were attached to the artillery reserve of the army of the Potomac, and on the February following, by direction of General Hunt, chief of artillery, army of the Potomac, the batteries were equipped for field service.

The muskets of the men were turned in and the light artillery organization assumed. The equipment was nearly that of light artillery, viz: one caisson to each piece, one battery wagon and forge to each battery, ten horses to each gun, eight horses to each caisson, six horses to each battery wagon and forge; ten per cent of spare or led horses; officers, sergeants and buglers mounted.

In drawing animals we were permitted to select large, heavy horses for the guns, which should always be per-

* For a time under Maj. T. S. Trumbull, 1st Connecticut artillery.

mitted, particularly for wheel horses. Light artillery harness was used, heavy breast hooks for the wheel horses being carefully selected; wheel traces were required for the swing horses on the guns, as they are attached to the lead bar. The implements for the piece were as follows:

Six handspikes; small ends under sweep-bar resting on axle, large ends resting on splinter bar, secured by leather straps from the hounds passing through loops on hand spikes to buckles on the forks.

One short roller on the stock between lunette bolts, secured by strap and buckle fastened to stock, and passing around the roller in its groove.

One trace rope; two half hitches in middle around the cascable, ends turned around the manœuvring bolts, and crossing to take up the length. This secures the piece from sliding on its carriage.

Sponge and rammer, (on the same stave,) lying on the piece, secured at each end by straps passing through loops on the staves and buckling around the cascable and chase. Spare sponge and rammer, carried same as sponge and rammer.

Wormer and ladle, one to each section.

Gunner's haversack, substituted for pass boxes and carried in limber chest of caisson.

The caissons were the same as for light artillery; having however only two partitions in each half chest, parallel and four and a half inches from each side—the outer spaces for projectiles, the inner for cartridges. This arrangement allowed forty-eight rounds to each caisson. They can be arranged for more, but on the march we had usually to transport forage on the caissons, and it was not thought best to carry more weight of ammunition. A tray in right half centre space of limber chest carried haversacks, pouches, primers and fuzes. One spare wheel and pole for caissons accompanied each section. Picket lines for each section were carried on spare wheel; these served for drag-ropes when required.

The battery wagon and forge were the same as for light batteries. About half of the spare parts for field carriages were turned in and a few spare spokes, linch-pins, nuts and washers for siege carriages drawn. Two spare pintles should be added. Two spare poles for siege carriages were carried on top of battery wagon, secured by strap and buckles. A spare siege limber should be attached to each battery to prevent the loss or delay of a piece by the disabling of a carriage.

THE BATTERIES ON THE MARCH.

The batteries were designated as siege batteries B and M, 1st Connecticut artillery. Battery B went into park with the artillery reserve, near the headquarters of the army of the Potomac. Battery M retained its position on Stafford Heights, opposite the left of Fredericksburg, occupied by it in the action at that place December, 1862. In the action with General Sedgwick's command at Fredericksburg, from April 29 to May 6, 1863, (the time of the Chancellorsville battle,) and from June 5 to 13, (the time of the last crossing at Fredericksburg,) battery M was actively engaged, giving our forces much aid and the enemy much annoyance.

On the morning of June 14 we marched in the movement of the army of the Potomac from Falmouth to intercept the enemy in Pennsylvania. Part of the horses of the battery were under harness during the entire night in a heavy rain, with details of men from the battery engaged in removing a 100-pounder Parrott* from Stafford Heights to Falmouth station. This duty prevented the battery from moving in its place with the artillery reserve, and obliged it to get into column near the rear of the army and encounter all the obstructions and delays from broken bridges, bad roads, &c., which the advance, one of much haste and confusion, had occasioned. We moved through Stafford Court House, Dumfries, crossed the Occoquan at Wolf Run shoals, (a bad ford with high precipitous banks, difficult for light artillery,) and reached Fairfax Court House at noon

* This gun was sent to Fredericksburg for a special purpose. It was assigned to company B, 1st Connecticut artillery and was mounted on June 12 on Stafford Heights. On June 13 the enemy opened upon our troops from a battery distant 3,000 yards from this gun, but were promptly silenced by a single 100-pounder incendiary shell, which fell directly into their battery. The recoil dismounted the 100-pounder gun, which was not remounted, being at once ordered to the rear on account of the movement of the army.—H. L. A.

on the 16th, a distance of fifty miles, marched in about as many hours, with frequent halts from delays to the column in our front. This was the first march of the battery; the horses and drivers were almost entirely new in their duties; the road was very bad for artillery, and much cut up, with many deep, muddy places, with corduroy broken and torn up, tripping the horses and clogging the wheels, with bridges gone and creeks to cross swollen by heavy rains, and having steep and slippery banks ; all these disadvantages combined made it one of the hardest marches ever accomplished by the battery, and fully tested and proved its ability to move with an army, and overcome ordinary obstacles.

In this and subsequent marches it was found that the caissons met more difficulties than the guns, which were frequently passed over places where light pieces and caissons were fast. The wheels of the siege carriages were much larger and wider, and the weight of the piece was so well distributed that they readily moved along over places where field-carriages found much difficulty in following. We frequently carried three or four hundred weight of forage on the limber of the piece.

The shoes furnished for the siege carriage did not operate well. The toggle in one instance slipping out dismounted the piece, and the wheel-driver narrowly escaped injury. The keys get bent and difficult to insert; and even when the whole worked as intended, the adjusting made too much delay in a long column. The shoes were, therefore, soon discarded, and a heavy rope secured by a ring to the stock was substituted. This rope, passed around the felloe with two or three turns and held by a cannonier walking by the side of the piece, was found to operate easily and safely. With good drivers and heavy wheel horses, it is rare that clogging is needed.

From Fairfax Court House we moved with the artillery reserve, by way of Edwards's ferry, over Sugar-loaf mountain, through Frederick city, to near Gettysburg ; where we were ordered to Westminster, Maryland, and went into position to resist an expected cavalry attack. That we were not in action at Gettysburg is much to be regretted as was expressed by General Meade, commanding the army, and by General Hunt, chief of artillery. That guns of this range and calibre would have made terrible havoc in the enemy's forming columns and lines, none can doubt. I afterwards learned that the dismounting of the gun mentioned above, by the slipping of the toggle in the lock-chain, had been witnessed by a general officer, who, not knowing the cause, received the impression that the guns were not easily managed. He so reported to General Meade, and it was feared that they might embarrass movements if placed in front at that time.

After the battle of Gettysburg we rejoined the artillery reserve at Frederick city, moved over South mountain to near Williamsport, returned, and by way of Middletown and Berlin recrossed the Potomac to Virginia. Thence we accompanied the army by way of Warrenton junction to Culpeper Court House, returned to Fairfax Court House, moved forward to Warrenton, and thence to Catlett's station. We were attached to General French's command in the movement to Kelly's ford on November 9; marching fourteen miles, and going into action at a gallop about 2 p. m. We were the second battery to open fire, and did fine execution. General Birney, in command of the advance column, personally told me that the comparative ease with which the position was taken was owing to the "accuracy and effect of the battery's fire, which he had never seen equalled;" and that he believed it saved our forces much loss of life. General French's chief of artillery expressed similar opinions.

We rejoined the artillery reserve near Rappahannock station, were in the movement to Mine Run by way of Germania ford, were engaged the last day on the left of General Sedgwick's command, and returned by way of Ely's ford. At this place, in going down a steep and deep cut on the right bank of the river, an accident occurred in the breaking of the pintle of the second piece, which had not been foreseen or provided for, and which occasioned a little delay. The limber of the first piece returned and took the delayed piece over the river, where *a new pintle was forged*, and the piece overtook the battery in about two hours. The night previous was very cold, and it is probable that the pintle full of frost and poor in quality had broken nearly off in the start. If the battery had been provided with a spare limber or pintle, the delay would have been trifling.

The batteries went into winter quarters near Brandy station on the following day, and remained there till April, when they were unhorsed and joined to their regiment for siege operations before Richmond.

The batteries were horsed a little more than a year. At the beginning, officers and men knew nothing of light artillery service except the handling of the piece. The care of horses, and the drill and management as light artillery had to be learned entire. During this time they were with the army of the Potomac in all its movements, marching over five hundred (500) miles, keeping well up in the column of march, taking the lead in their turn, moving thirty miles in a day, at times passing through swamps and rivers, and over mountains and as difficult roads as any campaign is likely to be conducted upon. The drag-ropes were used only on two or three occasions, and more on the caissons than on the guns. We drilled in most light battery and brigade battery movements, and were frequently reviewed in the same form with the batteries of the artillery reserve.

I consider it in every way practicable to take batteries of this kind* with a moving army, and from observation believe they are more readily moved than the 20-pounder Parrott batteries, whose weight is thrown on the rear axle.

---

* These guns—4.5-inch Ordnance pattern of 1861—weigh three thousand four hundred and fifty pounds. The gun carriage and limber complete, without implements, weigh three thousand seven hundred and forty-three pounds. The projectiles weigh about thirty pounds, and are fired with 3.5 pounds of mortar powder.—H. L. A.

It at first gave me much surprise, and always excited surprise and comment from others, that guns apparently so cumbrous could be so readily and rapidly moved. The efficiency of guns of this calibre in action you so well understand that it would be superfluous for me to mention it.

My delay and failure to put this in better shape I regret. I hope that it may come to you in time and convey to you the information you require.

Very respectfully,

F. A. PRATT,
Late Captain 1st Connecticut Artillery,
Commanding Battery M.

Brevet Major Gen. HENRY L. ABBOT,
Late Com'dg Siege Artillery, Armies operating against Richmond.

APPENDIX C.---SIEGE TRAIN OF 1864-'65.

Report of Major S. P. Hatfield, ordnance officer of train.

Organization of train. Ordnance records and reports. Supply of batteries. Defects developed in our ordnance. Fort Fisher expedition. Final operations of siege. Removal of our own and captured ordnance. Conclusion.

NEWARK, N. J., *July* 24, 1866.

GENERAL : In compliance with your request I have the honor to submit the following as my final report upon the ordnance department, siege artillery, during the campaign resulting in the surrender of Richmond, and upon the subsequent operations of removing captured ordnance.

ORGANIZATION OF THE TRAIN.

On April 20, 1864, I was informed by you that it had been determined to organize a siege train on the following basis.*

Six 100-pounder guns, five hundred rounds of ammunition per gun ; forty 4.5-inch or 30-pounder guns, one thousand rounds of ammunition per gun ; ten 10-inch siege mortars, six hundred rounds of ammunition per mortar ; twenty 8-inch siege mortars, six hundred rounds of ammunition per mortar. Ten 8-inch siege howitzers were subsequently added.

All of these were to be afloat, provided with at least two hundred rounds of ammunitton per gun, by May 6, 1864.

It was ordered that the train should be fitted out at the Washington arsenal ; accordingly, on the 21st, preliminary arrangements were made there for the work.

The following articles were selected as the necessary accompanying outfit for the above-mentioned guns, and for the service of the train :

Forty-four siege gun platforms ; thirty-three siege mortar platforms ; two travelling forges A, with selected stores ; two battery wagons D, with selected stores ; eight mortar wagons ; one hundred hand-barrows ; one hundred paulins ; six garrison gins ; two field gins ; four sling carts, (large ;) six sling carts, (hand ;) thirty thousand sand bags ; two thousand pounds of rope (assorted) and twine for cartridges ; five thousand pounds of bar iron, (assorted ;) five hundred pounds of steel, (assorted ;) five hundred pounds of nails, (assorted ;) five hundred pounds of candles ; twelve augers, (assorted ;) twelve 2-foot rules ; twenty-four dozen files, (assorted ;) two hundred felling axes ; two hundred caisson shovels ; two hundred picks ; one hundred globe lanterns ; fifteen cross-cut saws ; twenty-five hand saws, (assorted ;) six screw-jacks ; two thousand pounds of grease ; four long pieces of timber, 12 by 12, to make gangways.

In selecting stores for battery wagons and forges, only those were taken

---

* Subsequently largely increased ; see table of guns in position on April 2, 1865, on page 160.—H. L. A.

which were likely to be useful to the guns composing the siege train. For the spare parts of carriages and implements the Ordnance Manual furnishes a very good list.

On April 22 I began to load the stores on schooners—ordinary canal schooners, averaging one hundred and sixty-five tons burden. The vessels were supplied as they were needed until eleven were loaded. Previous to loading I had each vessel intended to carry siege guns, carefully stanchioned and skids placed across the vessel amidships. These skids were notched, so that when the guns were in their places (parallel to the keel) the chase of one lay by the re-enforce of the next, with the trunnions so disposed that the guns could not roll without being lifted from their beds. The parts of carriages were lowered into the fore and after hatches and there disposed regularly with the wheels on top. The mortars were dismounted, and with their beds laid on skids on the deck. The 100-pounder guns were placed in various parts of the deck; the chassis and carriages lowered into the hole. Each vessel carried with its guns a full supply of implements and at least two hundred rounds of ammunition, which were so disposed as to have the vessel in proper sailing trim. The implements were placed on the ammunition, each box marked with its contents and put in the most favorable position for use. Battery wagons, forges, sling carts, gins, &c., were placed on various vessels so as to be readily available. Most of the powder was put in one vessel, only keeping with the mortars a small supply.

It is important that all boxes containing fuzes, friction primers, and articles necessary for the service of the guns should be distinctly marked and placed so as to be easily found. Also, that a complete record be kept of the contents of each vessel. Too much care cannot be taken in these matters.

The following table shows the number of guns, with the material above mentioned, which it was found practicable to put in each vessel; each vessel, with its guns, held carriages, platforms, and about two hundred rounds of ammunition per gun:

Cargoes of schooners.

No. of vessels.	Cargo in each.
1	Six 100-pounder Parrott guns.
4	Ten 30-pounder, or 4.5-inch guns.
1	Ten 10-inch mortars (siege.)
1	Twenty 8-inch siege mortars, and twenty Coehorn mortars.
1	Ten 8-inch siege howitzers, mortar wagons, and extra stores.
1	Powder.
1	Ammunition, fixed and unfixed.
1	Miscellaneous stores.

With the facilities at the Washington arsenal this could be loaded in a week; but in this case, where so much material was brought together from all the arsenals in the country, it was not until May 12 that the loading was complete and the vessels were anchored in the stream near Alexandria ready for orders. Part of this delay was made by waiting for front pintle 100-pounder chassis,centre pintle having been first furnished by mistake.

To Major J. G. Benton and all the officers and employés at the Washington arsenal,and to Captain E.S.Allen, assistant quartermaster in charge of water transportation at Washington, I am much indebted for the kind interest displayed and the promptness with which every assistance was rendered.

On June 20 orders were received to send the train to City Point, Virginia, where it arrived on June 23. There I received orders to unload four 30-pounder Parrotts and four 8-inch mortars, with one hundred rounds of ammunition for each, together with other ammunition for guns already in battery. The disadvantage under which this was done established the fact that such operations should be carried on at some place separate from the army depot and devoted only to the siege train.

On June 24, you accordingly selected Broadway landing, on the south bank of the Appomattox river, four miles from City

Point, as the place offering the best advantages for this depot, there being sufficient depth of water, retirement, and a central position from which to supply the batteries then located along the Bermuda Hundreds front and at Petersburg.

On June 25 the first vessels were brought up and the depot established. There was at this time available for my use only one structure called a dock, very narrow and extending to low-water mark. Outside of this an empty barge was placed, making a floating dock. Here the timbers I had brought proved extremely valuable, as by them I was able to land guns without trusting to the worthless dock. I immediately built two more docks of crib-work, which answered very well for the time. Owing to the press of work I was obliged to use these docks through the summer, which considerably increased the mechanical difficulties of the work. As soon as possible I erected three very substantial docks, which made the work of loading and unloading stores very simple.

Where so many vessels* containing so dangerous material were gathered, great attention was paid to safety. The powder, cartridges in barrels, and fixed ammunition were from the establishment of the depot concentrated as much as possible in few vessels. The main powder vessel was moored at a considerable distance from the rest of the fleet. All laboratory work, filling cartridges and shells, was carefully conducted upon another vessel; the men working without shoes and the floor of the boat (covered with canvas) being frequently swept. This work was carried on without accident of any kind; although, from information received since the surrender of Richmond, we have learned that an attempt was made to blow up the vessels at the depot, as was done with the ordnance stores at City Point on August 9th, 1864. It was defeated owing to the vigilance of the guard and the disposition of the vessels.

*Twenty-one vessels and a tug in the autumn of 1864.—H. L. A.

The cost of the charter of vessels to contain the guns, ammunition, and other material in depot, and to receive the material in the batteries when necessary to embark it, was about ten thousand five hundred dollars per month.

The following table shows the batteries and armaments of each, on April 2, 1865, when operations ended:

Batteries and armaments of each in position April 2, 1865.

Designation of batteries.	Armament.																		Total guns in each battery.
	Field guns.								Siege guns and mortars.										
	Smooth-bore.				Rifled.				Smooth-bore.						Rifled.				
	12-pdr. light guns.	12-pdr. howitzer.	24-pdr. howitzer.	32-pdr. howitzer.	20-pdr. Parrott gun.	3-inch ordnance gun.	3-inch Parrott gun.	6-pdr. Sawyer gun.	8-inch S. howitzer.	Coehorn mortars.	8-inch S. mortars.	10-inch S. mortars.	10-inch S. C. mortars.	13-inch S. C. mortars.	30-pdr. Brooke gun.	30-pdr. Parrott gun.	4.5-inch siege gun.	100-pdr. Parrott gun.	
Fort Brady																3		4	7
Battery Sawyer													2					2	4
Battery Spofford																3		3	6
Bat'ry Wilcox and Parsons													1					3	4
Battery Drake					3		2												5
Battery Carpenter	2	2																	4
Battery Anderson					2											2			4
Battery Marshall	2					2													4
Battery, rear of Marshall							2												2
Battery McConihe	2																		2
Battery England	4					2													6
Battery Pruyn		2			2											1			5
Battery Dutton			1	2															3
Battery Burpee		4																	4
Battery Converse					2				2										4
Fort Cummings	1							1			2				1		1		6
Battery No. 4, Petersburg																3			3
Battery No. 5, Petersburg											4					3			7
Ft. McGilvery, Petersburg																1			1
Battery No. 8, Petersburg										2									2
Battery No. 9, Petersburg										3									3
Battery No. 10. Petersburg										3	4								7
Battery No. 12, Petersburg										8	2								10
Fort Morton, Petersburg												4					2		6
Fort Avery, Petersburg																4			4
Battery No. 20, Petersburg											4								4
Fort Davis, Petersburg																	2		2
Battery No. 2, City Point									2										2
Battery No. 5, City Point									2										2
Battery No. 7, City Point									2										2
Battery No. 8, City Point																	2		2
At depot									4	20	4	6	3	1	..	24	10	1	73
Total of each kind	11	8	1	2	9	4	4	1	12	36	20	10	6	1	1	44	17	13	200

ORDNANCE RESPONSIBILITY AND RECORDS.

The organization of a system of responsibility and battery reports was the first thing for attention. Some experience having already been gained in serving the siege artillery in the Bermuda Hundreds front, the following order was published:

[Orders No. 5.]

HEADQUARTERS SIEGE ARTILLERY,
In the field, June 26, 1864.

The following will govern responsibility for the ordnance property of the siege train:

I. Each company commander will give a memorandum receipt for his guns, implements, equipments, &c., to Captain Hatfield, and will be held responsible for their loss; articles lost being regularly invoiced subsequently. Unless so invoiced, no ordnance papers need be made out, except by Captain Hatfield.

II. A careful record will be kept of all firing, stating kind of ammunition used, number of each kind which took the grooves, number of each kind which burst well, &c. This record, or a paper stating that no firing took place, will be forwarded daily with the morning reports, together with a statement of ammunition received during preceding twenty-four hours, and the amount remaining on hand. Ammunition will be expended directly by Captain Hatfield upon this report.

By order of Colonel Abbot:

B. P. LEARNED,
First Lieut. and Acting Ass't Adjutant General.

The form of the morning report being thus left to the battery commanders, it was soon found that more uniformity was necessary. Accordingly I prepared a form, and Order No. 21 was published.

[Order No. 21.]

HEADQUARTERS SIEGE ARTILLERY,
Broadway Landing, Va., July 11, 1864.

I. The daily ordnance report from each battery called for by Order No. 5, from these headquarters, dated June 26th, will hereafter be made according to the form given below, for batteries where it can be used.

In the column for remarks will be noted the kind of ammunition used; number of each kind which took the grooves; number of each kind which burst well; ordnance stores required; kind of projectile preferred; in short, anything needed for the efficiency of the battery will be stated, and attention be called to anything which has occurred worth notice—these reports being the daily medium of communication between the batteries and these headquarters.

II. On the arrival of the ammunition supply train from these headquarters every evening, a commissioned officer of each battery will give his personal supervision to the unloading of the wagons, and will sign the ticket which each driver brings, noting any discrepancy.

III. Commanding officers of batteries will return at once by the empty wagons to these headquarters, any articles which have become unserviceable, or are useless in their battery; also all empty packing boxes, giving at the same time a list of articles so returned to the teamster of the wagon.

Morning report of battery.

* * * * * * * * * *

By order of Colonel Abbot:

B. P. LEARNED,
First Lieut. and Acting Ass't Adjutant General.

Battery commanders received still further instructions in Order No. 58:

[Order No. 58.]

HEADQUARTERS SIEGE ARTILLERY,
Broadway Landing, Virginia, September 18, 1864.

Hereafter, in each battery, a man will be detailed whose duty it is to note for each shot the information required in par. I, Order No. 21. From these notes the report will be compiled by the battery commander. The notes themselves will also be transmitted to these headquarters with the morning report.

By order of Colonel Abbot:

B. P. LEARNED,
First Lieut. and Acting Ass't Adjutant General.

From this time no difficulty was experienced in obtaining a record of the firing.

An addition was afterwards made to the form to show the number of shots fired by each gun, and another form suitable for mixed batteries. These forms, as finally fixed, are given below. They were printed by the quartermaster department at Bermuda Hundred.

Morning report of Battery , commanded by for the day of , 1864.

Armament.	Ammunition.																
	Guns.								Mortars.								
	Percussion shell.	Fuze shell.	Case shot.	Solid shot.	Canister.	Cartridges.	Friction primers.	Fuzes.	Shell.	Cannon powder.	Mortar powder.	Rifle powder.	Fuzes, wooden.	Fuze plugs.	Fuzes, paper.	Friction primers.	Quick match.
On hand last report																	
Received.																	
Total																	
Expended.																	
Remaining on hand																	

Rifled projectiles.				Fuzes.							
Kind of fuze used.	Took grooves.	Tumbled.	Uncertain.	Burned well.	Burned short.	Burned long.	Did not burn.	Uncertain.	No. of shots fired from each gun.	Ordnance No. of gun.	
Time											
Percussion											
Combination											
Wooden											
Paper											

NOTE.—In the column "burned well" will be noted all "time or percussion shells" which burned at time intended.

REMARKS.

———— ————,
Commanding Battery.

Morning report of battery , commanded by , for the day of , 1864.

Armament.	Guns.																			Mortars.														
	100-pdr.							30-pdr.												10-inch.				8-inch.				Coehorn.						
	Shells.	Case shot.	Solid shot.	Canister.	Percuss'n plugs.	Time plugs.	Cartridges.	Percuss'n shells.	Fuze shell.	Case shot.	Canister.	Cartridges.	Percuss'n shell.	Fuze shell.	Case shot.	Canister.	Cartridges.	Frict'n primers.	Paper fuzes.	Shell.	Mortar powder.	Rifle powder.	Wooden fuzes.	Shell.	Mortar powder.	Rifle powder.	Wooden fuzes.	Shell.	Mortar powder.	Rifle powder.	Fuze plugs.	Paper fuzes.	Quick match.	Slow match.
On hand last report																																		
Received																																		
Total																																		
Expended																																		
Total on hand																																		

	Percussion.	Fuze.	Case.	Combination.	Solid.	Percussion.	Fuze.	Case.	Combination.	Solid.	Percussion.	Fuze.	Case.	Combination.	Bormann.	Solid.	Wooden.	Wooden.	Wooden.	Paper.	No. of shots fired from either gun.	Ordnance number of gun, 100-pdr.	No. of shots fired from either gun.	Ordnance number of gun, 30-pdr.	No. of shots fired from either gun or mortar.	Ordnance number of gun or mortar.	Remarks.
Projectiles. Took grooves																											
Tumbled																											
Uncertain																											
Fuzes. Burst well																											
Burst short																											
Burst long																											
Did not burn																											
Uncertain																											

———————,
Commanding Battery.

These records of firing, checked by opening a sheet of issues and expenditures which was balanced daily at the depot to guard against mistakes, were transferred to abstract sheets of which one was prepared for each calibre in each battery. The following is an example of these sheets:

Fort Brady—Ordnance No. of guns, 11 *and* 15.

November, 1864.	Kind of—		No. fired.	Took grooves.	Uncertain.	Tumbled.	Burned well.	Burned short.	Burned long.	Did not burn.	Uncertain.	No. fired from 11.	No. fired from 15.	Remarks.
	Gun.	Projectile.												
1	100-pdr. Parrott..	Percussion shell..	10	8	1	1	7	2	..	1	..	5	5	
	100-pdr. Parrott..	Fuze shell	6	5	..	1	6	..	..	..	..	..	6	
2	100-pdr. Parrott..	Percussion shell..	4	4	..	..	3	..	..	1	..	4	..	
3	....	....	..	..	..	..	..	..	..	..	..	..	..	No firing.
4	100-pdr. Parrott..	Case shot.........	4	3	1	..	2	..	..	2	..	4	..	

From these abstracts were prepared the reports showing the number of shots fired by each kind of gun, in each battery, and the percentage of projectiles serviceable in regard to rifle motion and fuzes.

SUPPLYING BATTERIES.

This question also received early attention. As a first step, Colonel Abbot procured from Brigadier General Ingalls, chief quartermaster armies operating against Richmond, a train of fifty wagons; which it was estimated would be sufficient for the transportation, and which proved to be so for the ordinary operations of the siege. A certain number of rounds per gun was fixed for the supply of each battery, and from the morning reports was ascertained the amount of ammunition required. At first difficulties were experienced by the teamsters taking articles to the wrong batteries, notwithstanding efforts to direct them properly. Therefore each teamster was furnished with a ticket stating the contents of the wagon and which battery it was for.

This ticket after delivery was signed by some commissioned officer of the battery, and returned to the ordnance office. In addition to these tickets, a letter was conveyed by the wagon-master in charge of the train to the battalion commander, to whom the train reported, stating to which batteries ammunition was sent; and from his headquarters orderlies conducted the proper wagons to each battery. It was also found convenient to advise the battalion commander by telegraph when the wagons started, or if none were sent, so that the orderlies could be ready. This system proved perfect in insuring the proper supply of ammunition to the batteries.

At the depot First Lieutenant William C. Faxon had charge of the wagon train, and made up the list of battery supplies; being also responsible for the small-arms kept at the depot for issues to the troops. After his promotion, and since December 7, 1864, First Lieutenant J. H. Westervelt has performed these with the additional duty of most of the office work. First Lieutenant J. C. Gillett had direct charge of all laboratory work, also the arrangement and issues of all stores. Men selected from the regiment for their tact in these various branches, assisted him in these duties. A careful man was detailed to attend to collecting from the batteries all unserviceable articles which required the action of an inspector.

The allowance of ammunition to each gun in batteries varied to suit circumstances, those batteries in most prominent positions or having but few guns having the greatest supply; one hundred rounds being the maximum, not including canister. Thus, the supply was 100 rounds to each gun and howitzer, except Fort Brady, which had 75 rounds to each gun; battery No. 5, 8-inch mortars, 75 rounds each; battery No. 9, 24-pounder Coehorn mortars, 100 rounds each; battery No. 10, 8-inch mortars, 75 rounds each; battery No. 10, 24-pounder Coehorn mortars, 60 rounds each; battery No. 12, 24-pounder Coehorn mortars, 60 rounds each; battery No. 12, 8-inch mortars, 100 rounds each; Fort Morgan, 10-inch mortars, 80 rounds each; battery No. 20, 8-inch mortars, 100 rounds each; Redoubt Dutton, 24-pounder Coehorn mortars, 100 rounds each; Fort Brady, 8-inch mortars, 50 rounds each; Batteries Wilcox and Parsons, 10-inch mortars, 100 rounds each; Battery Sawyer, 10-inch mortars, 100 rounds each. This was found sufficient for all ordinary expenditures, and could be readily returned to depot if a hasty move became necessary.*

Owing to the unequal number of guns in the batteries, the supply of implements (except where one article is furnished to each gun) proportioned to the guns, as laid down

---

* As an illustration of the speed with which such transfers can be made the following instance is cited:

At 11.35 p. m., July 30, 1864, I received a telegram from General Hunt to move with urgent haste certain siege ordnance and siege materiel from the batteries at Petersburg to the depot at Broadway landing. The following table exhibits what was moved. The distances are accurately taken from the engineer maps of the army of the Potomac. It will be seen that the mean distance per piece, exactly computed, is eight miles.

Name of battery.	Armament.	Distance to depot.	Distance for one piece.	Remarks.
		Miles.	*Miles.*	
Fort Sedgwick	6 4.5-inch guns	9.3	55.8	5th corps, front.
Near Battery 20	6 8-inch mortars	9.1	54.6	Do.
Near Fort Rice	10 10-inch mortars	8.5	85.0	Do.
Battery 17	6 4.5-inch guns	8.0	48.0	Do.
Near Battery 18	6 Coehorns	8.8	52.8	Do.
Fort Morton	6 4.5-inch guns	8.0	48.0	9th corps, front.
Near Battery 14	4 8-inch mortars	8.2	32.8	Do.
Battery 4	3 30-pdr. Parrotts	5.4	16.2	18th corps, front.
Battery 1	4 30-pdr. Parrotts	4.0	16.0	Do.
On railroad, near Battery 3	1 13-inch mortar	......	....	Sent six m. to City Point.
Total	52	......	409.2	

The seven siege guns, &c., in batteries 1 and 4 were moved by transportation—light artillery teams and wagons—furnished by Colonel Piper, chief of artillery 18th Corps.

The 13-inch mortar, which was served on a railroad truck car made so strongly as to resist the shock of firing, was drawn to City Point by a locomotive. The rest of the materiel was moved by the four artillery teams of Captain Korte and by a train furnished by General Ingalls, as follows: My regular train, 50 wagons; an extra train of 60 wagons, furnished for contingencies arising from

in the ordnance manual, has not been found sufficient. Allowance should be made for batteries of two guns in estimating the required implements for a siege train.

From the arrival of the train at City Point, on June 23, 1864, to April 3, 1865, sixteen hundred and fourteen wagon-loads of material were sent out to the batteries, one ton being an average load.

The following table shows the total expenditure of ammunition by each kind of gun in each battery during the entire siege:*

---

the battle, upon my requisition on July 30th; and an extra train of 60 wagons and 18 eight-mule teams, furnished at 8.30 a. m., July 31st, in response to a telegram of mine, dated 1.35 a. m. of that date; total, 170 wagons and 22 teams.

The orders to move the materiel were received by me at 11.35 p. m., July 30th. By the aid of the telegraph matters were so well arranged that the trains began arriving at the depot at daylight of July 31st, and continued to do so as fast as they could be unloaded, up to 2.30 a. m. of August 1st, when the last was received; total period occupied in removal, 27 hours.

The materiel was all brought in government wagons, except the guns and the 10-inch mortars, the latter of which were loaded on mortar wagons. The confederates did not discover the movement, although many of the batteries were in the very front of our line. The aggregate weight transported was 225 tons.

The materiel was shipped as fast as unloaded. By noon of August 1st, 36 hours after the first telegram, everything was afloat.

The labor at the depot was performed by two companies of the 4th New York artillery and six companies of the 37th New Jersey volunteers, the latter working two at a time, aided occasionally by the companies of the 1st Connecticut artillery who had served and moved the batteries. Three wharves were used, which were at this date mere crib gangways.

Everything was brought away—artillery, ammunition, implements, platforms, mantelets; nothing was damaged or lost. Much of the unusual promptness of this movement was due to the facilities furnished by the telegraph.—H. L. A.

* The following is the number of shots fired in the principal battles of this period by the siege artillery brigade:

Petersburg mine........................ 3,833 shots, weighing about 75 tons.
Fort Steadman .... 1,060 shots, weighing about 16 tons.
Final battle........................... 5,560 shots, weighing about 85 tons.

H. L. A.

Total expenditure of ammunition during siege.

Designation of battery.	Armament.																	Total number of rounds fired from each battery.
	Field guns.								Siege guns and mortars.									
	Smooth bore.				Rifled.				Smooth bore.					Rifled.				
	12-pdr. light.	12-pdr. howitzer.	24-pdr. howitzer.	32-pdr. howitzer.	20-pdr. Parrott.	3-in. Parrott.	3-in. Ordnance.	6-pdr. Sawyer.	24-pdr. C. mortar.	8-in. S. mortar.	10-in. S. mortar.	10-in. S. C. mortar.	13-in. S. C. mortar.	24-pdr. Sawyer.	30-pdr. Parrott.	4. 5-in. Ordnance.	100-pdr. Parrott.	
Battery near Fort Brady															145			145
Fort Brady										87					164	899	206	1356
Dutch Gap	277							196	614	2379							182	3648
Battery Sawyer												579		10			480	1069
Battery Spofford					24										242	632	224	1122
Battery Wilcox and Parsons												301					468	769
Battery Drake					202	40				20								262
Battery Carpenter	54	71																125
Battery Perry					16													16
Battery Anderson					720										652			1372
Battery Marshall	105						155											260
Battery McConihe	42																	42
Battery England	209						141											350
Battery Pruyn		1			374										310			685
Battery Dutton			164	514					42									720
Battery Burpee		40																40
Fort Converse					393										234			627
Battery No. 1, Petersburg															833			833
Battery No. 4, Petersburg															4018			4018
Battery No. 5, Petersburg										6662					4801			11463
Battery No. 8, Petersburg									680									680
Battery No. 9, Petersburg									4165									4165
Battery No. 10, Petersburg									415	4661								5076
Battery near No. 11, Petersburg									271	1023								1294

Total expenditure of ammunition during siege—Continued.

Designation of battery.	Armament.																	Total number of rounds fired from each battery.
	Field guns.								Siege guns and mortars.									
	Smooth bore.				Rifled.				Smooth bore.					Rifled.				
	12-pdr. light.	12-pdr. howitzer.	24-pdr. howitzer.	32-pdr. howitzer.	20-pdr. Parrott.	3-in. Parrott.	3-in. Ordnance.	6-pdr. Sawyer.	24-pdr. C. mortar.	8-in. S. mortar.	10-in. S. mortar.	10-in. S. C. mortar.	13-in. S. C. mortar.	24-pdr. Sawyer.	30-pdr. Parrott.	4. 5-in. Ordnance.	100-pdr. Parrott.	
Battery No. 12, Petersburg									10198	3311								13509
Battery No. 14, Petersburg										836								836
Battery No. 15, Petersburg									315									315
Fort Morton											1609					1431		3040
Battery No. 17, Petersburg																1611		1611
Fort Avery															640	439		1079
Battery No. 20, Petersburg										2096								2096
Fort Rice											360							360
Fort Sedgwick										14						375		389
Fort Davis																180		180
Near Fort Walsh															170			170
13-inch mortar battery													218					218
Total	687	112	164	514	1729	40	296	196	16700	21089	1969	880	218	10	12209	5567	1560	63940

DEFECTS DEVELOPED IN OUR ORDNANCE.

The record of firing shows the quality of ammunition.*

* * * * * * * *

In the 4.5-inch Schenkl sabots a wide difference was observed in quality; only that having the mark of the Washington arsenal, dated 1864, worked well. An examination showed that in this the sabots had sufficient fibrous strength to give the rotary motion to the projectile, while the other sabots were extremely brittle.

Eight thousand new sabots were furnished, which gave good results except in the case shot, which still failed frequently. The reason for this was not explained, as they were similar to other shells in shape, and position of the centre of gravity in relation to the figure.

The Schenkl combination fuze was not of much service, as it could not be made to burn over eight seconds. Beyond 'that time the heat seemed to be so great as to melt the metal of the fuze plug. I think a harder metal should be used in making these fuzes.

Nearly all the Parrott projectiles were those having the composition ring. To render them more certain to take the grooves, the ring was always slightly started with a cold chisel, each battery being provided with one for that purpose.

A few shells captured from the rebels, of their own make, were used in our 30-pounder and 100-pounder guns.

In the 30-pounder and 4.5-inch gun the cartridges used were of mortar powder, weighing 3.25 pounds.

In the mortar batteries, the powder used was from the mills of Dupont, Hazzard, and the Oriental Company. All seemed to give about equal results. Before firing, the powder was well mixed in the barrels to secure uniformity of range.

The reports show that about 70,000 friction primers were

---

* These tables having been given on pages 42 and 99, are not repeated.—H. L. A.

expended during the entire siege; that is, in firing about 64,000 rounds.

All of the 4.5-inch guns used showed that they would not endure much firing without being bouched. Seven of them became worthless by enlargement of the vent. Six nearly new guns were bouched before being subjected to steady firing in the battery. Only two guns of the siege artillery burst. A 5.82 inch Sawyer gun in battery Sawyer burst after firing a few times. It is not known how many times it had been fired previous to use here. One 30-pounder Parrott gun had about twelve inches blown off at the muzzle, by the explosion of a percussion shell in the bore. The gun was trimmed off and did not seem to suffer in accuracy or range.

A 100-pounder Parrott gun was spiked by breaking a gimlet in the vent. It was necessary to put in a new vent, as the old one could not be cleared or taken out, except by drilling. No other Parrott guns were rebouched, and with all our firing they showed but comparatively little wear.

To the 100-pounder gun batteries I sent shells unloaded and with both time and percussion fuzes, to use either as they were required.

I would recommend, as an improvement to mortars, that a line of metal be marked on them before sending them into the field. The Coehorn mortars too require this.

The carriages and mortar beds worked well throughout. The new model mortar beds should have some arrangement to hold the lanyard, in order to pull the friction primer in the right direction.

The long elevating screw attached to the cascable of the 30-pounder Parrott proved too weak, invariably breaking off the shoulder at the bottom of the screw where it fastens into the fork. In most cases, before breaking, the thread slipped from the bottom upward, until the end of the screw struck the bolt in the stock, thus allowing considerable play to the screw. The old style of elevating screw is much preferred by those who have used these guns.

During the heavy firing of April 2, a battery of six new 30-pounders, near the Avery House, broke all their elevating screws, after very little firing. Being without preponderance, the screws were taken off, and the gunners reported that the guns were subsequently handled much more easily during a time when the greatest activity was required.

No trouble was experienced by breaking of bolts in iron carriages.

In the new model platform for 100-pounder guns, the pintle should be keyed in, as it is apt to spring out from the shock of recoil on the chassis. This occurred twice in our batteries, causing considerable delay and trouble. Two guns in battery Spofford were thus temporarily disabled.

In the rifled guns no trouble was experienced by the grooves becoming foul, bristle sponges used occasionally having kept them clean.

The damage to our siege guns and carriages resulting from the enemy's fire consisted in the breaking of three wheels and the end of an axle outside of a wheel. The top carriage for a 100-pounder gun was hit fairly on the front of one of the cheek plates, totally destroying it. The gun detachment standing near it was not injured, except by very small splinters of iron. A 100-pounder gun, No. 11, in Fort Brady was also hit fairly in the face of the piece by a 7-inch rifle shot, producing three small fractures radiating from the axis of the piece. The gun was used without further development of these cracks. Many of the guns and mortars were struck by fragments of shells without being seriously injured.

The large sling carts always gave trouble when taking them in new roads through the woods, on account of the length of the pole making it impossible to turn short. Several poles were broken in this way, by the wheels striking stumps or trees. I therefore shortened them three feet, which much improved them for use. In transporting 100-pounder guns, three of these sling carts were broken off at the end of the axle body. The wood of which they were made was unfit for the purpose, being what is called "brash." Mortar wagons were often used instead of sling carts, and seemed to operate much better in carrying these guns a long distance.

The hand sling carts, when breaking, have given away in the axles at the shoulder of the journal, it breaking off without warning. I think if this were rounded, instead of having the present sharp angle, they would be improved.

In other respects, the implements furnished by the Ordnance department have answered every purpose for which they were intended.

For the outfit of a siege train, a small lathe that will cut threads, and a drilling machine, should be included; also a hammer for a pile driver, when the train depends upon vessels to constitute a depot.

THE FORT FISHER EXPEDITION.

On January 5th, 1865, I received orders to prepare for embarking sixteen 30-pounder Parrott guns, and twenty Coehorn mortars with implements, and five hundred rounds of ammunition per gun, to go with an expedition then ready to leave Fort Monroe. On January 7th, at eight a. m., the screw steamer C. C. Leary reported, and the loading immediately commenced. At five o'clock a. m. January 9th, the ordnance and stores were all on the vessel. The steamer registered about eight hundred tons, being about two hundred feet long and thirty feet beam. The lower hold being occupied with coal, most of the articles were put on the main deck. Four guns, one carriage, and a small amount of ammunition were put on the coal under the forward hatch. Five hundred rounds per gun and three hundred rounds per mortar were stowed five tiers of boxes high, occupying two-thirds of the deck, beginning to load at the stern. Forward of this, near the main hatch, were twelve guns laid on deck, securely wedged with carriages, platforms, planks, wheels, and ammunition boxes. These and miscellaneous stores, including six wagons, filled completely the whole space forward of the engine. The gun implements were put on the ammunition in the after part of the boat.

The cartridges and mortar powder were put in a small space under the main deck near the stern, where being surrounded with ballast, they were in the place of greatest safety from fire.

Besides the implements necessary to serve the guns, I had tools for blacksmith and carpenter work, and for erecting batteries; also a battery wagon, sling carts large and small, heavy and light tackle blocks, with a large supply of rope, assorted sizes.

While going down the James river, the cargo was examined and provisions made against accident which might be occasioned by any rolling of the vessel.

On January 13th we arrived off Fort Fisher, North Carolina. On January 15th, three 30-pounder guns with carriages, one hundred rounds of ammunition per gun, a large and small sling cart, and all implements necessary for the service of the guns, were landed.

For this work a large navy launch was used; the steamer being anchored, and a warp of three-inch rope, one hundred and twenty fathoms long, carried from her to the beach. The guns were lifted from the hold by strong tackles from the mast-head, and hauled out over the launch by another tackle from the end of the yard. Small guys in various directions prevented swinging of the guns as the vessel rolled. Each gun was laid across the launch on skids, and the launch was drawn along the warp as near the beach as possible. A strong hawser was then fastened to the gun, which was rolled overboard and hauled out of the water by about two hundred men on the shore.

The landing of the above-mentioned articles occupied the whole day, and may be taken as a fair example of the amount of work which can be performed under similar circumstances with a *smooth sea.* A slight roughness renders the operation impossible.

Acting Master Z. L. Tanner, United States steamer Rhode Island, superintended the landing, and the manner in which this difficult work was performed reflects the highest credit upon this officer.

The capture of the fort rendered further work unnecessary.* As the material landed could not be re-embarked from the ocean beach, it was turned over to Captain C. R. Bannon, company B, 1st Connecticut artillery, and the steamer C. C. Leary was sent to Beaufort to await further orders, Lieutenant L. W. Jackson being left in charge.

On February 9th, the ordnance and stores were returned to Broadway Landing; the last of the artillery detachment from the siege artillery brigade followed shortly after.

FINAL OPERATIONS OF THE SIEGE.

During April 1st, some artillery firing took place along the lines.

Early on April 2d, firing began through the whole line,

* The capture of Fort Fisher will doubtless be added to the list of operations wherein land batteries are claimed to have been subdued by guns afloat. While the fire of the fleet alone would have been powerless to cause the surrender of the fort, it is freely admitted that it silenced the barbette batteries, of which the confederate armament exclusively consisted, and enabled our land forces to dispense with the tedious operations of a siege in preparing for the assault,—an assistance of the most important character. The views already given on page 45 illustrate how, in my judgment, this effective service of the fleet might have been prevented by proper preparation on the part of the confederates. A knowledge of the actual armament of the two belligerents is important in studying this battle, and the following lists have therefore been prepared. That of the confederate ordnance was made by myself with the greatest care immediately after the capture; that of the fleet was furnished to me from the bureau of ordnance, Navy Department. It should be added, in explanation of the former, that about eight of the *carriages* of the guns reported in good order on the *land front* were disabled. The efficiency of not more than one of the guns on the *water front* was in the least impaired by the fleet, either by direct impact of shot, or by the destruction of carriages, or by the piling of sand in or around the muzzles, or in any other way; the only gun found disabled on this front was a 7-inch Brooke rifle which had evidently burst in service. This exemption from injury on the water front was doubtless caused by the fact that these guns were not much served;—probably because the heavy fire of the fleet rendered it inexpedient to attempt it, especially as most of these guns could not be traversed sufficiently to bear upon the only vessels which were really

and calls on the depot for ammunition became very urgent. Fifty extra wagons were procured from Captain Strang, assistant quartermaster at City Point. There were at

weakening the work by destroying its land defences. Want of ammunition had nothing to do with the confederate silence, as immense supplies of ordnance stores were captured.

Armament of Fort Fisher.

Position.	Armament.	Smooth-bore.		Rifled.		Total.
		Good order.	Disabled.	Good order.	Disabled.	
Land front	10-inch columbiad		1			1
	8-inch columbiad	4	1			5
	32-pdr. gun	3	2			5
	24-pdr. gun		1			1
	12-pdr. gun	2				2
	Coehorn mortar	2				2
	6. 4-inch rifle			4	3	7
	4. 2-inch rifle			1		1
	Total	11	5	5	3	24
Water front	10-inch columbiad	9				9
	8-inch columbiad	4				4
	8-inch rifle			2		2
	7-inch rifle			1	1	2
	6. 4-inch rifle			5		5
	Total	13		8	1	22
Fort Buchanan	11-inch columbiad	2				2
	10-inch columbiad	2				2
	12-pdr. field	1				1
	Total	5				5
In depot	10-inch columbiad	2	1			3
	10-inch mortar	1				1
	8-inch columbiad	3				3
	32-pdr. gun		2			2
	32-pdr. carronade	5	1			6
	24-pdr. gun	1				1
	24-pdr. Coehorn	2				2
	12-pdr. gun	2				2
	6-pdr. gun	4				4
	1. 5-inch gun		1			1
	Volley gun		1			1
	7-inch rifle				1	1
	6. 4-inch rifle			2		2
	5. 8-inch rifle			1		1
	4. 6-inch rifle			1		1
	4. 2-inch rifle				2	2
	3-inch rifle			3		3
	2. 2-inch rifle			4		4
	Total	20	6	11	3	40
	Grand total	49	11	24	7	91

Of these guns, those in Fort Buchanan and those in depot could not be directed against the fleet, and should not be considered in estimating its serviceable

Broadway Landing about one hundred and fifty men for work; these and the one hundred teams were kept busy; something over one hundred loads were sent out; this was all that number of men and teams could do in one day. The ammunition had to be lifted out of vessels' holds, car-

---

armament. It is evident that there were only twelve guns in position of any value against an iron-clad vessel; and most of them could not be brought to bear upon the monitors.

To fully understand the naval attack, the diagram contained in the appendix to the report of the Secretary of the Navy for 1865 should be examined. It represents the approximate position of the iron-clads, of the three "lines," and of the reserves. The following table exhibits the exact armament of those grand divisions of the fleet, arranged in two columns; one showing the entire number of guns, the other the number available in the bombardment—including all the pivot guns and half the broadside guns of the vessels in action, and excluding the bronze 12-pounders and 24-pounders as too small to be of use in such service.

The contrast between the two armaments is striking, and shows that without mortar fire or heavy iron casemates the land guns had little chance of success. The reports of Admiral Porter, however, state that their fire inflicted considerable damage in the different bombardments.

Armament of United States fleet in attacks on Fort Fisher.

Ordnance.	Iron-clads.		1st line.		2d line.		3d line.		Reserves.		Entire fleet.	
	Total armament.	Available in the bombardment.	Total armament.	Available in the bombardment.	Total armament.	Available in the bombardment.	Total armament.	Available in the bombardment.	Total armament.	Available in the bombardment.	Total armament.	Available in the bombardment.
SMOOTH-BORE.												
XV-inch	10	10	....	......	....	......	....	......	....	......	10	10
XI-inch	14	7	3	3	6	6	4	4	....	......	27	20
X-inch	....	......	....	......	1	1	1	1	....	......	2	2
IX-inch	....	......	48	25	189	95	25	14	....	......	262	134
VIII-inch	....	......	....	......	16	8	20	10	4	......	40	18
32 pdrs	....	......	6	3	....	......	26	13	2	......	34	16
24-pdrs. bronze	....	......	24	......	6	......	10	......	28	......	68	......
12-pdrs. bronze	....	......	15	......	23	......	8	......	12	......	58	......
RIFLED GUNS,												
200-pdr. Parrott, 8 inch	2	2	1	1	5	5	....	......	....	......	8	8
100-pdr. Parrott, 6.4 inch	....	......	10	7	13	10	8	7	1	......	32	24
60-pdr. Parrott, 5.3 inch	....	......	2	1	1	1	....	......	....	......	3	2
50-pdr. Dahlgren, 5.1 inch	2	1	....	......	....	......	1	1	....	......	3	2
30-pdr. Parrott, 4.2 inch	....	......	8	7	12	9	15	11	14	......	49	27
20-pdr. Parrott, 3.67 inch	....	......	8	5	1	1	6	6	6	......	21	12
Total	28	20	125	52	273	136	124	67	67	......	617	275

H. L. A.

ried up the dock, and loaded on the wagons. As the teamsters did not know the way to the batteries, a train of twenty or more wagons had to be made up before they could start; this occasioned considerable delay in supplying the batteries. On such occasions, officers commanding batteries should be very careful to state which kind of ammunition will most likely be wanted, and the Ordnance Office should be informed of the points where ammunition is being most rapidly expended. I experienced some embarrassment from the want of this; and the orders I received early in the day were too general to prevent considerable unnecessary work. A portion of the wagons were employed carrying ammunition to batteries where it was not needed, and batteries most needing it narrowly escaped exhaustion at critical periods of the day. This confusion was mostly owing to the sudden and unexpected removal of the telegraph operator at our post on April 1st. No accident occurred to any of our guns on April 2d.

REMOVAL OF OUR OWN AND CAPTURED ORDNANCE.

On September 29th, 1864, at the battle of Chafin's Farm, the 18th Corps captured and turned over to me the following articles, all of which, excepting the 4.2-inch Brooke rifled gun, were forwarded to Fort Monroe arsenal: One 8-inch columbiad; one 4.6-inch Brooke rifled gun; one 32-pounder gun, banded and rifled; eight 6-pounder field guns; one 30-pounder caisson; one 6-pounder caisson; one 32-pounder carronade. One 4.2-inch Brooke rifled gun fell through the pontoon bridge while crossing James river. It was raised by Lieutenant Gillett, and placed in position at Fort Cummings.

On April 3d, 1865, the work of removing our own and the captured batteries to Broadway Landing began. To this place were brought all the batteries of position in front of Petersburg, and on the line of the Appomattox, and the light guns along the right bank of James river to Fort Drewry.

On April 8th, Major Brooker, with his battalion of the 1st Connecticut artillery, was ordered to Chafin's Bluff to establish a depot for shipment of the heavy guns and material on that side of James river, Lieutenant Gillett, assistant ordnance officer, being in charge of the work. A few days after, Major Cook established a similar depot at battery Semmes, near the mouth of Proctor's creek. Lieutenant J. P. Elliott was appointed assistant ordnance officer here.

On May 1st, the loading of ordnance at Broadway Landing was completed, and the main depot moved to Drewry's Bluff. At this place were received all the guns above battery Semmes.

The following table* gives the kind and calibre of the captured guns shipped at the various depots, and shows the point where they were taken.

---

* It must be remembered that this table exhibits only the artillery removed by the siege artillery brigade. Vast numbers of captured field and siege guns and much ammunition were removed by the different chiefs of light artillery from the vicinity of Petersburg, and especially from the lines of Richmond proper.—H. L. A.

List of captured ordnance removed by siege artillery brigade.

Removed from—	Smooth bore.																					Rifled.															Total.
	12-pdr. Coehorn mortars, (iron.)	24-pdr. Coehorn mortars, (iron.)	8-inch siege mortars.	8-inch sea-coast mortars.	10-inch sea-coast mortars.	6-pdr. light guns, (iron.)	6-pdr. light guns, (brass.)	12-pdr. light guns, (iron.)	12-pdr. howitzers, (iron.)	12-pdr. mountain howitzers, (brass.)	12-pdr. howitzers, (brass.)	24-pdr. howitzers, (brass.)	32-pdr. guns, (smooth and short.)	32-pdr. guns, navy, (smooth.)	32-pdr. guns, (smooth and long.)*	8-inch guns, (smooth, banded, short.)†	8-inch siege howitzers.	8-inch columbiads.	10-inch columbiads.	9-inch Dahlgrens.	8-inch Brooke guns, (bored to 10-inch smooth.)‡	3-inch guns.	3.67-inch Blakeley guns.	3.67-inch Parrott guns.	4.2-inch Parrott guns.	4.2-inch Brooke guns.	4.5-inch Dahlgren gun. §	4.6-inch Brooke gun.	6.4-inch navy gun, (banded.)	6.4-inch sea-coast gun, (banded.)	6.4-inch Brooke gun.	7-inch Brooke gun.	Columbiad bored to 5.8 inches. ‖	8-inch Brooke gun.	8-inch columbiad. ¶	8-inch Parrott gun.	
Lines of Petersburg	4	16	7	..	..	..	..	..	..	..	..	..	..	..	..	..	2	1	..	..	....	..	..	..	..	..	..	..	..	..	..	..	..	..	..	..	30
Lines of the Appomattox	..	..	2	..	..	2	..	..	..	1	..	2	2	..	..	..	..	1	..	..	....	1	1	1	1	4	1	1	..	2	..	..	..	..	..	..	22
Chester station, on railroad	..	..	..	..	..	..	..	..	..	..	1	..	..	..	..	..	..	..	..	..	....	..	..	..	..	..	..	..	..	..	..	1	..	..	..	..	2
Lines of Bermuda Hundreds	..	..	..	..	..	..	..	..	..	1	..	..	..	..	..	..	..	..	..	..	....	2	..	..	..	..	..	..	..	..	..	..	..	..	..	..	3
Light guns between Dantzler and Fort Drewry	..	2	..	..	..	3	..	2	1	..	..	..	..	..	..	..	..	..	..	..	....	3	..	..	..	..	..	..	..	..	..	..	..	..	..	..	11
Water batteries Dantzler and Semmes	..	..	2	1	3	..	..	..	..	..	..	..	..	..	..	..	..	..	8	..	....	..	..	..	..	..	..	..	..	..	..	2	..	1	1	..	18
Water battery Semmes to Fort Drewry	..	..	..	..	2	..	..	..	..	..	..	..	..	..	..	..	..	2	..	..	2	..	..	..	..	..	..	..	..	1	..	3	1	1	..	..	12
Fort Drewry	..	..	..	..	..	..	..	..	..	..	..	..	2	..	3	..	..	3	6	..	....	..	..	..	..	..	..	..	..	..	..	1	..	..	..	..	15
Land batteries in rear of Chafin's	4	11	1	..	..	9	6	2	1	..	..	..	1	2	..	4	..	..	..	..	....	1	..	..	..	..	..	1	1	2	..	..	..	..	..	..	46
Water batteries at Chafin's Bluff	..	..	..	..	..	..	..	..	..	..	..	..	..	..	..	..	..	2	12	..	....	..	..	..	..	..	..	..	..	..	1	..	..	..	..	1	16
Wreck of C. S. gunboat Drewry	..	..	..	..	..	..	..	..	..	..	..	..	..	..	..	..	..	..	..	1	....	..	..	..	..	..	..	..	..	..	..	..	..	..	..	..	1
Total	8	29	12	1	5	14	6	4	2	2	1	2	5	2	3	4	2	9	26	1	2	7	1	1	1	4	1	2	1	5	1	7	1	2	1	1	176

* Somewhat different from 32-pdr. sea-coast gun, old style.
† 8-inch guns (resembling 4.6-inch Brooke) on siege carriage.
§ Gun cast and banded for 8-inch rifle, but bored for 10-inch smooth.
‡ U. S. navy gun, cast without trunnions, having Dahlgren brass breech straps.
‖ Gun cast somewhat heavier than 8-inch columbiad; shape like it. Bored 5.8-inch and rifled.
¶ Gun cast somewhat heavier than 10-inch columbiad; shape like it. Bored 8-inch and rifled.

For the removal of these heavy guns from batteries, there were used one large sling cart,* found at the Tredegar works, (wheels ten feet diameter,) and our own large sling carts and mortar wagons. The 8-inch rifles were moved by the large cart; 10-inch columbiads and 7-inch rifles by fastening two of our sling carts together. Mortar wagons carried 8-inch columbiads, 32-pound banded guns, and the mortars.† For the heavier loads the motive power was chiefly men. The guns were dismounted by removing the "counter hurters," and running the carriage off on skids placed behind the chassis on the prolongation of its rails. Ropes were then attached and the gun and carriage upset sideways by the men. No damage to the carriages could be detected. At Chafin's Bluff and Depot Semmes, strong, flat lighters were used for floating docks. At the former place the guns were rolled on skids to their position on the vessel's deck; at Depot Semmes, shears were rigged, and the guns hoisted from the dock over the vessel's rails, and lowered to their place on deck. I think this the easier way to load such guns on such vessels (with high rails) as were supplied to us for transportation. At Drewry's Bluff the loading was done on barges, without rails, from the ordinary wharves at that place, Captain J. M. Twiss superintending the work.

On July 9th, 1865, the embarking of the captured material was finished, and the labor of the 1st Connecticut artillery with the siege artillery of the armies operating against Richmond was done.

---

* This cart was designed for the two 12-inch guns cast at the Tredegar works just before the evacuation. They were cast solid and bored out, weighing nearly 50,000 pounds. This cart, when carrying an 8-inch rifle or heavy columbiad over the bad roads, near the batteries, was moved by twelve mules and one hundred and fifty men, using drag ropes—H. L. A.

† The chassis of the larger guns being too wide to sling under our sling carts, a pair of long skids were attached to the axles of the cart and its limber, forming a long-geared wagon, which easily carried them. The top carriages were removed on mortar wagons.—H. L. A.

CONCLUSION.

I have presented many details in this report, hoping that in case another siege train should be put in the field, the officer who may have charge of it will have some basis to work on, and will see at the start the chief objects aimed at, which nothing but such practical experience as we have had will define. Of course, the details will vary according to circumstances, and must be managed to suit the occasion by those having the work in charge. I have not yet found that anything more could have been done in recording the results of such operations as ours, and I trust that they may be found useful in showing the relative value of guns and ammunition used, by giving the test of actual work, carefully observed, on a scale which will probably not occur again.

I desire to express my appreciation of the zeal and ability displayed by my assistants, Lieutenants Faxon, Gillett, and Westervelt, in all the operations connected with the service of the siege train; also, to Lieutenant Jackson for his eminent services in embarking the siege train at the Washington arsenal, and at the time of the Fort Fisher expedition. To them, and to Captain Twiss and Lieutenants Elliott and Loomis, who performed excellent services, with the officers and men who have been on duty at the ordnance depot, I must express my thanks for cordial support at all times given, without which my labors must have been much more difficult.

To yourself, General, I am grateful for the confidence and appreciation shown me ; and although the great work in which we were associated is finished, the recollections of it will always be among the most pleasant of my life.

Very respectfully, your obedient servant,

S. P. HATFIELD,

Late Major 1st Connecticut Artillery,

Acting Ordnance Officer Siege Artillery.

Brevet Major General H. L. ABBOT,

Late Commanding Siege Artillery before Richmond.

DESCRIPTIVE LIST OF PLATES

Plate I. FUSES AND MISCELLANEOUS DIAGRAMS.

Fig. 1. Confederate polygonal cavity
2. Confederate polygonal cavity
3. Confederate powder measure for Coehorn mortars
4. Maury, 10-inch
5. Maury, 8-inch
6. Maury, 8-inch
7. Smoothbore, test shot for proving barrels
8. Smoothbore, 10-inch for armor crushing
9. McEvoy "attachment" fuse igniter
10. Armstrong combination fuse
11. Tice concussion fuse
12. Schenkl combination fuse
13. Sawyer combination fuse
14. Broun (Clay?) percussion fuse, "b" should rest on plate below domed cap and "a" should extend below the weight.
15. Armstrong percussion fuse
16. Plane table, 15-inch gun
17. Plane table, 15-inch gun
18. Plane table, 15-inch gun
19. Cast of vent after firing
20. Field work
21. Field work
22. Adams hand grenade fuse
23. McIntire concussion fuse
24. Confederate 4.2-inch siege rifle
25. Confederate 4.62-inch siege rifle
26. 15-inch gun experiments
27. 15-inch gun experiments

Plate II. UNITED STATES SIEGE AND FIELD RIFLE PROJECTILES, 1864.

Fig. 1. Parrott, 100-pdr., 6.4-inch (long)
2. Parrott, 100-pdr., 6.4-inch, incendiary, double cavity
3. Parrott, 100-pdr., 6.4-inch (short)
4. Parrott, 100-pdr., 6.4-inch, chilled-nose bolt
5. Sawyer, 5.82-inch, flanged, Navy
6. Schenkl, 4.5-inch siege rifle
7. Absterdam, 4.5-inch siege rifle
8. Dyer, bolt, 4.5-inch siege rifle
9. Dyer, shell, 4.5-inch siege rifle
10. Parrott, 30-pdr., shell
11. Parrott, 30-pdr., incendiary, double cavity
12. Schenkl, 30-pdr., shell
13. Sawyer, 3.67-inch, shell
14. Hotchkiss, 20-pdr.
15. Parrott, 20-pdr.
16. Schenkl, 20-pdr. (long)
17. Schenkl, 3-inch, shell
18. Schenkl, 3-inch, case
19. Hotchkiss, 3-inch, shell
20. Parrott, 3-inch, shell

Plate III. CONFEDERATE 8-INCH RIFLE PROJECTILES.

Fig. 1. Read-Parrott, probably U.S.
2. Tredegar Parrott
3. Read-Parrott, bolt
4. bolt, unidentified sabot pattern
5. "Tennessee sabot", armor punching
6. Brooke, bolt
7. Brooke, shell
8. Brooke, bolt, milled base
9. Armstrong, bolt
10. Blakely, bolt
11. Blakely, shell, flanged
12. Blakely, shot, flanged
13. Armstrong, copper studs

Plate IV. CONFEDERATE 7-INCH RIFLE PROJECTILES.

Fig. 14. Read, shell
15. "Tennessee sabot", shell
16. "Tennessee sabot", bolt
17. "Tennessee sabot", bolt (short, flat nose)
18. Brooke, shell
19. Brooke, bolt
20. Brooke, shell
21. Brooke, shell
22. Brooke, shell
23. Brooke, armor piercing shell
24. Brooke, shell
25. Dahlgren pattern shell
26. Whitworth, shell
27. Brooke, bolt, milled base
28. Brooke, bolt, milled base, armor punching

Plate V. CONFEDERATE 6.4-INCH RIFLE PROJECTILES.

Fig. 29. Read, shell
30. Read, shell
31. Dahlgren pattern shell
32. "Tennessee sabot", shell
33. "Tennessee sabot", shell
34. Read pattern converted to "Tennessee sabot", shell
35. Read pattern converted to "Tennessee sabot", shell
36. "Tennessee sabot", shell
37. "Tennessee sabot", bolt
38. experimental armor punching
39. experimental armor punching
40. experimental armor punching
41. Brooke fragment, base and sabot
42. Brooke, bolt, milled base
43. Dahlgren pattern shell
44. Dahlgren pattern shell
45. Brooke, bolt, milled base

Plate VI. CONFEDERATE SIEGE AND FIELD RIFLE PROJECTILES.

Fig. 46. "Tennessee sabot", shell
47. Hotchkiss, shell, for 50-pdr. Dahlgren rifle, U.S.
48. Read, shell, rifled 12-pdr.
49. "Tennessee sabot", shell, rifled 12-pdr.
50. "Tennessee sabot", shell, rifled 12-pdr.
51. Brooke, bolt, milled base, rifled 12-pdr.
52. "Tennessee sabot", bolt, rifled 12-pdr.
53. Read-Parrott, shell, 4.2-inch
54. Read-Parrott, shell, 4.2-inch
55. Read-Parrott, shell, 4.2-inch
56. "Improved" Read-Parrott, shell, 4.2-inch
57. Read-Parrott, shell, 4.2-inch
58. Read-Parrott pattern
59. Brooke, shell, 4-2-inch
60. Brooke, shell, 4.2-inch
61. Schenkl, shell, 4.2-inch, U.S.
62. Read-Parrott, shell, 3.5-inch
63. Read-Parrott, shell, 3.67-inch (20-pdr.)
64. Read-Parrott, bolt, 3.67-inch
65. Blakely, shell
66. Dahlgren, hollow bolt, 3.4-inch, U.S.
67. Blakely, 3.5-inch, probably segmented case shot
68. Hotchkiss, 3.4-inch, U.S.
69. Hotchkiss, shell, 3-inch, U.S.
70. Armstrong, segmented case shot, studs, English
71. Armstrong, shell, studs, English
72. Armstrong, shell, 3-inch, lead jacket
73. Whitworth, shell, 12-pdr.
74. Whitworth, shell, 12-pdr.
75. Whitworth, bolt, 12-pdr.
76. Archer, bolt, 3-inch
77. Burton, case, 3-inch
78. Read, shell, 3-inch
79. Read, shell, 3-inch
80. Read, shell, 3-inch
81. Read, shell, 3-inch
82. Read, shell, 3-inch
83. Read, shell, 3-inch
84. Read-Broun, shell, 3-inch
85. Whitworth, bolt, 6-pdr.
86. Whitworth, shell, 6-pdr.
87. "Tennessee sabot", shell, 2.25-inch mountain rifle
88. probably Ellsworth, bolt, 1.5-inch, U.S.